John Lindsay

Practice in Cotton-Carding

A complete manual for the card room of the cotton mill

Arthur Young

Peace and Reform, Against War and Corruption

PEACE AND *REFORM,*

AGAINST

WAR AND CORRUPTION.

IN

ANSWER TO A PAMPHLET,

WRITTEN

BY ARTHUR YOUNG, ESQ. ENTITLED,

" The EXAMPLE of FRANCE, A WARNING to BRITAIN."

" *I shall always have the Satisfaction to have aimed sincerely*
" *at Truth and Usefulness, though in one of the meanest Ways.*"
LOCKE.

LONDON:

PRINTED FOR J. RIDGWAY, YORK STREET, ST. JAMES'S
SQUARE.

MDCCXCIV.

PEACE AND REFORM,

WAR AND CORRUPTION.

AN attentive perufal of many political
Pamphlets, produced in the beginning of
the prefent year, and an opinion that it would
not be difficult to expofe the fallacy of the arguments upon which were founded the moft popular of thofe written againft the Caufe of Freedom,
led me firft to think of attempting their refutation ; and fome leifure during the autumn
months, enabled me to try what I had previoufly
wifhed to have feen done by any abler hand.
I have efpoufed that fide which is, for the moment, the leaft popular ; and obfcure as both
this Pamphlet and its Author may be, I am therefore prepared to expect fome fmall fhare of that
obloquy, which is now fo largely beftowed on
all thofe who prefume to queftion the wifdom
of the meafures of government. Secured from
danger by power, and uncontradicted by reafon,
becaufe of the danger, the advocates of Corrup-

tion

tion have of late exulted, almoſt unoppoſed, in their triumph. The moſt moderate Favourers of Reform have been blended with the moſt inſane Zealots for Revolution ; Toleration and Athoiſm ; Peace and Regicide, have, by the ſupporters of abuſes, been wickedly deemed, and by the multitude weakly believed, to be ſynonimous terms.

The national underſtanding thus miſled and prejudiced ; the temper of a very great majority of the People rendered furious and vindictive ; the partizans and the partakers of Corruption bold, active, and cruel ; the current has been, from neceſſity, ſuffered of late to run wholly in favour of the moſt abſolute Toryiſm. Few men would enter on a labour ſo perilous and unacceptable as that of expoſing the Deluſion. To incur odium from the many for the approbation of the already approving few ; to write cramped up as much as if pinioned in the Pillory, leſt it ſhould actually be the reward ; to oppoſe reaſon to paſſion, and be certain of being unſucceſsful, whether right or wrong, were conſiderations ſufficient to intimidate Prudence by the danger of the attempt, and by its hopeleſsneſs, even to ſilence the moſt honeſt Zeal in the cauſe of Freedom.

Accordingly the Preſs laſt winter was much more occupied by Tory or High Church and King opinions, than two years before it had been by thoſe of an oppoſite deſcription. Their abundance was even greater than the harveſt which followed the labours of Mr. Burke ; and although their arguments may be futile and ridiculous, they cannot fully be anſwered, while

New-

Newgate and the Pillory are called in to support them.

The firft Pamphlet that fuggefted to me the idea of endeavouring to expofe thefe exploded, inconfiftent, mifchievous, Doctrines, was, that of Mr. Arthur Young. Several reafons induced me to think it a proper object of animadverfion ; for, although on the firft perufal it appeared to be fuch a jumble of contradictions, falfehoods, and even libels on that Conftitution which it profeffed to defend, that I did not believe common fenfe could endure it ; yet on reading the approbation of Mr. Reeves, and finding it had been circulated with great induftry, I concluded it might be of fome ufe to refute what I thought fo fufceptible of refutation ; and while the apparent eafinefs of the tafk feemed to fet its accomplifhment within my reach, an object fo inglorious did not promife to provoke a Rival. Mr. Reeves has, by his approbation, adopted its opinions as his own, and, however worthlefs it may be in itfelf, it no doubt derives fome confequence from being thus promulgated as the Manifefto of his Committee.

Its defign is to deter us from making a Parliamentary Reform, by exciting our horror at the atrocities which have taken place in France ; thefe atrocities Mr. Young falfely afcribes to the principles of Liberty ; and afferts, that Englifhmen would be equally guilty with their neighbours, if once they began political improvement. With *Reform* he connects *Property* ; and attempts to eftablifh as a fact, that a government more purely reprefentative than our own is at prefent, could not long exift without an Agrarian Law ;

thus

thus, deceitfully alarming all men of wealth and advising them to join in the war against the French, as the only means of preferving whatever they hold dear. To follow him regularly through every falsehood and abfurdity would neither be entertaining nor useful. I shall therefore begin with fome specimens of his unfairnefs: I shall shew that the crimes of France, fo far from having any natural connection with her principles, are the very fame which the rage of Faction has led the Friends of War and Corruption in England, if not to perpetrate, at leaft to recommend in fupport of Principles directly oppofite; and which, if admitted to be proof againft the principles of French Liberty, muft alfo be proof againft the principles of Mr. Young, and the other English Tories, as the blind advocates of each fyftem equally applaud them. After having difcuffed thefe preliminary matters, I shall proceed to the confideration of the two great queftions of Reform and Peace, againft which Mr. Young and his fellow labourers have raifed fo many prejudices.

His difingenuoufnefs is evident at the commencement of his book, the very foundation of which is laid on palpable mifreprefentation. He owns himfelf to have been a "warm friend" to the firft Revolution, yet the chief part of his induftry is employed in condemning Mr. Paine, Major Cartwright, &c. for their writings in defence of it. He reprobates them for having done what he himfelf did, yet he does not own he did wrong. Had he like others repented his miftake and made his recantation, he then, with a better grace, might have attacked thofe with whom

whom he formerly had agreed; and, like **Mr.** Pitt in the laſt debate on Parliamentary Reform, might have maintained, that conſiſtency was a proof of a want of judgment, and that it was always to be preſumed thoſe were in an error who did not change their opinion.

" The Revolution before the 10th of Au- " guſt" (ſays Mr. Young), " was as different " from the Revolution after that day as light " from darkneſs; as clearly diſtinct in prin- " ciple and practice as Liberty and Slavery. " The ſame principles which directed me to " approve the Revolution in its commence- " ment, the principles of *real Liberty*, led me " to deteſt it after the 10th of Auguſt." Here he aſſerts his approbation of the Revolution *up to* the 10th of Auguſt, which he acknow- ledges was conducted, till that period, on the principles of *real Liberty*; and he alſo ſays in the ſame page, " How little reaſon therefore " to reproach me with ſentiments *contrary* to " thoſe I publiſhed before the 10th of Auguſt;— " I am *not* changeable, but ſteady and con- " ſiſtent."

Compare this with page 21, where, in ſpeak- ing of the Revolution, he affirms, " it has " brought more miſery, poverty, devaſtation, " impriſonment, bloodſhed and ruin, on France, " in *four years*, than the old government " did in a century."—If the " *principles* and " *practice* of the Revolution up to the 10th of " Auguſt were conformable with real liberty," how could they have brought miſery, poverty, &c. on France for *four years*? How comes it that he approves the devaſtation, bloodſhed, and

and ruin *before* the 10th of Auguſt and yet diſ-
approves of them afterwards ?—But it is in fact
the old Deſpotiſm he contends for, which (p. 33)
he calls " the mildeſt and moſt benignant go-
" vernment in Europe, our own only excepted;
" a government cruelly libelled in the cha-
" racter given by one of our reforming
" Orators."—Mr. Young, however, libels it
ſtill more in his own Travels, publiſhed long
after the Revolution, where, ſpeaking of the natural
richneſs of the country, he ſays " the diſpenſa-
" tions of Providence ſeem to have permitted
" the human race to exiſt only as the prey of
" Tyrants, as it has made pidgeons for the prey
" of hawks."—" Oh! if I were a legiſlator of
" France, I would make ſuch great lords ſkip
" again!"—Yet this is the government which
he now calls " regular and mild." *—I would
not quote theſe paſſages, if Mr. Young had
owned his approbation of the firſt Revolution
had been miſplaced : But, on the contrary, he
affirms he is " ſteady and conſiſtent."

Mr. Paine's works he treats as if they had
been written and publiſhed ſince the 10th of
Auguſt, 1792. The panegyrics contained in
" The Rights of Man" on the Conſtitution
of 1789,—A conſtitution of which Mr. Young
declares his unaltered approbation, are falſely
and artfully tortured into panegyrics on the
Convention, and the events that have taken
place ſince the deſtruction of monarchy. For
inſtance, he gives part of a ſpeech of Marat in
January 1793, wherein the Convention is called

* P. 49.

" a ſcandalous

" a fcandalous fpectacle—an affembly of mad-
" men and furies," and immediately follows it
by obferving, that " Paine is of a contrary
" opinion, he faid they debate in the language
" of gentlemen ; their dignity is ferene, &c."
Thefe paffages in Paine were publifhed long
before the 10th of Auguft; and with equal
juftice might the eulogiums of Mr. Burke, on
the character of the French people, prior to
1789, be quoted as eulogiums on the maffacres
of the fecond of September.

But Mr. Young's reafonings are not more
inconfiftent with the opinions he pretends to
entertain, than they are with each other, for
" The example of France, a warning to Britain"
may not improperly be called " the example
" of France an example to Britain." It re-
proaches the French with deftroying the
Liberty of the Prefs, and advifes the very fame
thing to be done in England ; it condemns
the principle of arming one clafs of men againft
another, and foon afterwards fets forth the neceffity of following the example here, by arming
the rich againft the poor ; and the violation
of the freedom of election at Paris is held up
to derifion, while corruption and prefcriptive
election in England are applauded as the true
foundation of national profperity. Indeed he
is not fingular in thefe blunders, for bloodfhed
and rapine would be equally practifed in this
country as in France, if we may take the *will*
for the *deed.* Here the theory has been inftilled
and admired ; there the practice has been
adopted. And the only difference between the
French and Englifh Marats,—a difference which
does

does not diminifh their guilt,—is, that the one would affaffinate all who deny, and the other would affaffinate all who maintain the fovereignty of the people.

The faults of the Convention, of public bodies, and even of individuals, are carefully collected and detailed as the deliberate acts not only of thofe who rule in France, but of the whole nation. The moft falfe affertion or the moft wicked counfel, though coming from an infignificant fool, or difregarded madman, is fufficient to excite Mr. Young's execration, and to criminate a whole people. How little lefs guilty can the reafonable part of mankind think this kingdom, if their judgment is formed in the fame man-ner. Crimes equally deteftable have been preached and vindicated in this country, by perfons of eminent ftation; and if the fanguinary fcenes of Paris have not been repeated in Lon-don, Manchefter, and Birmingham, it is not for the want of Britifh Marats, and Roberfpierre's to inftigate and protect the inftruments. If affaffination is to be committed, it is indifferent to me who are the objects, or what are the mo-tives;—whether the victim be a plunderer of the poor induftrious Swinifh Multitude, or of pam-pered tyrannical Nobles, and lazy mifchievous Priefts. The offence is as heinous in the one cafe as in the other; and juftice might be as much violated, were Marat or Roberfpierre, as if Windham or Burke were to be murdered with impunity.

As a companion to Mr. Young's fhocking picture of France, I fhall give the outline of a propofed one in England. But if my colouring is not fo bold as his, it muft be accounted for by the peculiar circumftances under which I write, rather than in the weaknefs of my fubject. What he fays of Chabot, Roberfpierre, or Danton, it may be dangerous to fay of a prieft, a counfellor, or a judge in England. He might decorate his tales with the moft glaring falfehood and rancorous calumny, and be in unifon with the nation and in favour with the government. The difapprobation of the firft, and refentment of the fecond, might be my lot, were I to indulge equally in much better grounded invective. His reafoning is enforced by aids which it would be imprudent, probably dangerous, for me to ufe ; for when it is difficult to demonftrate his argument, or fuftain his affertion, he attacks humanity with a pike, a dagger, or a guillotine, and, addreffing himfelf to the feelings, inftead of the underftanding, triumphantly exults in his abfurdity and falfehood.

If " Marat's grand fpecific of cutting off " * 150,000 heads" be compared with Mr. Reeves's " excefs of virtue to exterminate the " Diffenters,†" which fhall, we think, moft criminal ? There are more than 150,000 diffenting heads in Britain, therefore Mr. Reeves outdoes Marat ; but, on the other hand, we muft recollect that he does it in a more delicate manner. Marat fpoke roundly out ; Mr. Reeves only *infinuates*. Marat's, however, were words,

* Young p. 69.
† Vide the *original* publication of Thomas Bull to John Bull, owned and apologized for by Mr. Reeves ; and alfo the Comments of Mr. Fox in his Speech, Dec. 13, 1792.

and

and might have been the ebullition of the moment, but Mr. Reeves's was a deliberate act,
which, from its nature, muft have been confidered and approved by a body of men, the
Crown and Anchor Committee. In whatever
way we compare thefe two circumftances, we
fhall find them nearly on a par; but their
effects in this country were very different.
Marat's advice has been the theme of horror in
every company, though it did not extend to
them; on the contrary, Mr. Reeves's has been
little noticed, though there are very few perfons who have not fome friends among the
diffenters, and confequently were affected by
it. The more we examine the characters of
thefe two gentlemen, the greater fimilarity we
fhall find between them. How far they have
been juftly rewarded, I fhall not pretend to
decide : Mr. Reeves, befides enjoying about
half a dozen places under government, has
lately been paid between three and four thoufand pounds—Marat has been affaffinated.

Mr. Reeves, in his " reflections on the pre-
" fent crifis," fays, " it would be well if thofe
" who diflike the Englifh Conftitution would
" remove to another region." What decree
of the Convention would occafion emigration
more certainly than this fentiment, if it had
the power? And Judge Afhhurft, in his charge
to the grand jury on the 19th of November,
1792, at the very moment the decree of fraternity was paffing in France, declared, " there
" have, however, under the beft fyftems of go-
" vernment, been found men of corrupt prin-
" ciples, who, having forfaken honeft induftry,
with

" wifh to throw every thing into confufion, and
" to live by rapine and plunder ; when that is
" the cafe, it is become neceffary for the coer-
" cive power of the ftate to lend its reftrain-
" ing hand, and to punifh offences of fuch a
" flagrant nature. There is no profpect of
" reformation, till fuch *corrupt members* be
" *cut off* *." The wide fenfe which would be
given to the laft expreffion by the inflamed
populace, among whom this charge was liberally
diftributed, it is not very difficult to imagine.

The Faft Day, inftead of being paffed in conformity with its profeffed purpofe, in humiliation
before God, in prayers for the converfion of
unbelievers, the reformation of ourfelves, and
the general peace and happinefs of mankind ;
inftead of a day on which every prieft made an
extraordinary exertion of his powers in imploring
the benevolence of the Almighty to enlighten
the minds, to foften the hearts, and to fpare the
blood of his people, it was chiefly celebrated by
the moft dreadful maledictions. The Supreme
Being, who, true religion tells us, enjoins bro-
therly love, forgivenefs, humanity and virtue,
was addreffed by our divines as if he had been
more mercilefs and blood-thirfty than any divinity
that ever difgraced paganifm ; and the temples of
the God of Peace were made to refound with
imprecations, from which even our anceftors
would have recoiled when engaged in the wor-
fhip of their ferocious Odin, whom they revered
as " the terrible and fevere God; the active
" roaring deity ; the father of flaughter ; the

---

* Vide the minifterial Gazette, the Times of November
20th, always very correct in its law reports.

" God

" God that carrieth defolation and fire, and
" nameth thofe that are to be flain *."

The folemnity of the fcene was well calculated for roufing and mifleading the paffions,
and every artifice was employed to excite hatred
towards the French, and provoke us to fury. The
priefthood as well as the princes felt themfelves
interefted in the caufe, and their zeal fhook
the pulpit with exhortations to vengeance. The
Bifhop of Gloucefter, before the Houfe of Lords,
thus fpoke of that nation ;—" Infatuated and
" remorfelefs people ! The meafure of your
" iniquities feems at length to be full ; the hour
" of retribution is coming faft upon you ! Drunk
" with the blood of your fellow citizens, you
" have dared to fpread your ravages abroad ;
" roufing the furrounding nations, in juftice to
" themfelves and the common caufe of hu-
" manity, to confederate againft you, *in order*
" *to execute the wrath of God on your devoted*
" *heads.*" His lordfhip, however, might have
been reftrained from fuch rafh denunciations of
divine judgment, by the awful admonitions of
the founder of that religion which he pretended
to preach."

" And Jefus anfwering, faid, fuppofe ye thefe
" Galileans were finners above all the Galileans,
" becaufe they fuffered fuch things ? I tell you
" nay,—But except ye repent, ye fhall all like-
" wife perifh."

" And thofe eighteen on whom the Tower of
" Siloam fell and flew them, think ye that they
" were finners above all men that dwelt in Je-

* See the Edda.

" rufalem ? I tell you nay : but except ye re-
" pent, ye fhall all likewife perifh*."

The other Faft Day Sermons were in unifon
with that of the bifhop of Gloucefter, with a
very few exceptions. The Rev. J. Gardener,
at Taunton, faid, " Shall we not labour to
" bring fuch perfons" (as the French, and Reformers in general) " to a proper fenfe of their
" duty, *or exterminate them and their opi-*
" *nions?*" and the Rev Mr. Bromeley, at Fitzroy
Chapel, hopes " that the reckoning which God
" will make will not be long delayed againft a
" nation," (France) " which is certainly be-
" hind no other whofe meafure of iniquities
" has in any records of time *called forth his*
" *vengeance to erafe it from the earth* †."
Thefe are the fentiments of our high church
paftors : Such is the religion, the benevolence,
the humanity, they teach ! To exterminate for
opinion ! What more did Marat ever defire !
To be the inftrument of God in executing his
vengeance, Mahomet ufed the fame plea for
all his murders and rapine ! To erafe a whole
nation from the earth ! ! ! Neither Mahomet,
Marat, nor Roberfpierre, have equalled this !
How limited and infignificant have been their
profcriptions compared with thofe of our own
pious paftors, who would " feal on the fore-
" head as the fervants of God ‡," all thofe who
make war againft France ; who would " fend

---

* St. Luke, c. xiii.

† Thefe paffages are taken from the fermons publifhed
under the names of thofe divines.

‡ Vide Revelations, c. vii. v. 3.

" myriads

" myriads of locufts, with crowns like gold upon
" their heads, and faces like men, invefted with
" fcorpion power, to torture the unfealed" enthufiafts of that diftracted nation, and " let
" loofe the angels of the Euphrates to flaughter
" a third of mankind * !"

Similar paffages from the fermons preached
on that *Chriftian* day would fill a volume.
Moft of them tend to inflame the people to a
war of extermination, and infinuate the deftruction of thofe who defire a Parliamentary
Reform. Surely our divines cannot be fo much
miftaken as to imagine thefe harangues gratifying to the Head of the Church? Their
affection towards the crown, indeed, is natural. The Bifhop of Durham's promotion
has taught them the road to preferment ; and
my Lord of Gloucefter has been long looking
for a tranflation : but not fuch as Elijah's :
His prefent ambition looks no higher than
Canterbury.

Nor were the fermons publicly preached
more inflammatory than the writings anoni-
moufly publifhed by our High-Church Men ;
one of which, in Birmingham, under the
fictitious name of Job Nott, thus fpeaks of
thofe whom he calls " New-fafhioned, reftlefs
" Diffenters," and the members of a fociety
inftituted on the principles of Mr. Pitt and the
Duke of Richmond for procuring a Reform of
Parliament. " Do be off—only think of the
" New Drop—you may be recorded in the
" Newgate Calendar—tranfportation may re-
" form you—*you deferve to be highly exalted—*

* Vide Revelations, c. ix,

" Did

" Did you ever fee the New Drop ?" and concludes with wishing that these Dissenters and
Reformers, whom he deems factious, " tied in
" their garters may swing." Yet this elegant
author calls himself a friend to conciliation and
unanimity, a *moderate* man, a man of *peace !*
He may be so for a *Birmingham* man ; but if
such are the friends to peace and moderation in
that town, can we wonder at the atrocities
which have taken place there, and still may be
repeated, while Job Nott and such publications
are publicly sold with a bookseller's name to
them, and are even boasted of by their authors * ?

Members of both Houses of Parliament have
acted in unison with the church, and have also
repeatedly insisted that a war of extermination is the only thing which can save this
country. The defence of the riot at Man-
chester equals the exhortations of the pulpit.
We not only find riots indirectly recommended,
but openly vindicated. Unlawful violence
against a man for his political opinions is justified in the British House of Commons ! Here the
" Sacred duty of insurrection" is preached as
unequivocally as ever it was in France, where
actual and excessive danger is in some degree a
palliation of criminality. But in England no justi-
fication of similar atrocities could be shewn, except in the unfounded apprehensions artfully excited by a false and pernicious alarm.

---

* One very curious article contained in Job Nott, de-
signed probably to give confidence to the Church and King
Partizans, is, " that Sir Robert Lawley never is kept wait-
" ing by Mr. Pitt, *when* he sends in word it is on *Bir-*
" *mingham business.*"

Mr.

Mr. Reeves's Committee detected in their outfet, infinuating it to be an excefs of virtue to exterminate the Diffenters, did not alter their courfe, though they more carefully concealed their proceedings. The ftreets were over-run with the moft violent inflammatory libels; the war-whoop againft Diffenters and Reformers was fung at every corner; and, if affaffination was not committed, it was not for want of prompters. Some of the hand-bills and ballads in circulation laft winter would even have fhocked Mr. Burke! And it is fortunate that the lower had more difcretion and humanity than the higher ranks of perfons who encouraged fuch publications, otherwife we might have had a Tory *Second* of *September* in England; and the bloody anniverfary, a diftinguifhed *red* letter day in the " Church and King" Calendar, might have been celebrated by our Bigots.

Not one of the authors or diftributors of thofe incendiary papers was even feized, much lefs punifhed: but a poor bill-fticker, who could not read, was tried and imprifoned for pofting up bills in favour of Parliamentary Reform; and Holt, the printer, at Newark, has been convicted of a libel for re-printing a paper on the fame fubject, which was firft circulated under the aufpices of Mr. Pitt and the Duke of Richmond, when they contended for that meafure; and which it is not improbable was originally written by his Grace, as it ftrongly contends for Univerfal Reprefentation.

I fhall not comment on what Mr. Young repeatedly infinuates, though he does not plainly defire

defire, by fuch paffages as " The King of
" France died on the fcaffold, becaufe he would
" not fhed the blood of Traitors, Confpirators
" and Rebels ; he liftened to thofe who peti-
" tioned for Reform ;" and I fhall alfo pafs over
the many oral threats of affaffination made by
individuals, the anonymous letters to the fame
effect, one of which at Staines, threatened the
houfe and life of Mr. Fox, and conclude this
flight fketch, of the defire of fome men, to fee
" Church and King" maffacres in this country,
by giving as an epitome of the others, an extract
from a hand-bill, circulated at Exeter, in December laft. " Our noble King hath made a
" fine fpeech from the throne to his Parliament,
" as muft be acknowledged by every well-wifher
" to his country, and as for them that do not
" like that and the prefent CONSTITUTION, let
" them have their deferts, that is a HALTER
" and a GIBBET, and be burnt afterwards, not
" as PAINE hath been, in effigy, but in body
" and perfon. To which every loyal heart will
" fay Amen !"
Thefe facts, I think, are fufficient to juftify me
in afferting, that the affaffinations, and other
crimes in France, are not detefted by certain
perfons in this country on account of their
enormity, but that thofe who execrate moft
bitterly, only wifh to imitate them on a different
clafs of victims. Marat's fall was regretted, by
our Tories, becaufe he laid the eggs of mifchief;
and the death of the King of France, was declared
in the Houfe of Commons, by a member of Adminiftration, to be, in fome refpects, a fortunate

D   circumftance

circumstance in rousing the British Nation to War.

The animating soul of Mr. Young's book, being the atrocities of the French, which he falsely ascribes to their principles, I have thought it necessary at my commencement to shew, that, according to his own reasoning, he condemns his own principles, since the same atrocities have been recommended by the advocates for his opinions; and, that if I might indulge in the same furious execration, I could find cruel doctrines enough in England to balance against him:—to mislead and inflame those who swallow assertion for fact, and invective for argument. I do not, however, defend the crimes of the French, although I think, as far as crimes can be excused, no people in the world ever had more to plead in extenuation, because no people in the world ever were so irritated by internal treachery, and alarmed by external danger.

But, proscription and cruelty, are not the only parts of the conduct of the French which have been imitated in this country. We had our Tory Jacobin Associators at the St. Alban's Tavern, and our more furious and degraded Club of Cordeliers in the Crown and Anchor Committee. If in Paris " Persons have been imprisoned by order of individuals," others, at home, have been voted in a state of accusation, upon the authority of anonymous letters transmitted to Mr. Reeves. The affiliated Jacobin Clubs of the French Provinces have been made the model of the Reevesian Associations and Committees against Republicans and Levellers at home; and in the English

English Parochial, as well as in the French Popular Meetings, the Conſtitution, as declared by Reeves, Pitt, and Co. has been formally approved. Every Preſident or Chairman of theſe Conventions becomes a Municipal Inquiſitor: Sir Joſeph Banks makes his *viſites domiciliaires* in the pariſh of St. Anne, and keeps a regiſter of the complexion, age, employment, &c. of lodgers and ſtrangers. The ſection of St. James's. *denounce* for *incivifm* every houſe-keeper who does not oblige his ſervants, workmen, and apprentices, to ſign their acceptance of the Conſtitution. No tradeſmen is to be employed, who has not been fraternized by the officers of his diſtrict; no publican is to be licenſed, who has not reported *fuſpected perſons.* Every man is called upon, more palpably than in France, to declare our Conſtitution glorious and unreformable; and if any one is more conſiſtent than Mr. Pitt and the Duke of Richmond, and ſignifies a wiſh for the removal of abuſes in the conſtruction of the Houſe of Commons, he is branded as a Jacobin, and if poſſible, utterly ruined. Is not this governing by the very ſame means, ſo much execrated in the French: by mobs, by terror, by popular coercion? The great maſs of the people, from the higheſt to the loweſt ranks, are ſummoned to conſtitute themſelves into partial arbitrary tribunals to acquit and to condemn.

The dread and regret with which the inſtitution of theſe Engliſh Primary Aſſemblies was beheld by the rational Friends of Freedom, found ſome conſolation in the hope that they would not be of long duration. It was imagined when

the

the temporary alarm was paſt, that they would not only be difufed, but the public would condemn the delufion, and give the government of the country to thoſe who conſtitutionally ſhould poſſeſs it, and whoſe reſponſibility is a check upon their intemperance, the King's Miniſters. But inſtead of this, we find our Tory affiliated Societies, converted into the inſtruments of the Church and King Jacobins, and executing without authority or reſponſibility, thoſe meaſures which adminiſtration deſire, but cannot with decency tranſact. They denounce all who wiſh to petition the King or Parliament, contrary to the ſentiments of the Placemen and Penſioners at the Crown and Anchor Tavern; the inhabitants of Glaſgow have been threatened with Church and King vengeance for preſuming to complain of that which is daily ruining them,— the War; and our Parochial Clubs are knocking down the Conſtitution, in imitation of the Clubs in France, by voting contributions to ſupport the army. Every perſon may ſoon be obliged, under pain of denunciation as a traiterous Jacobin, to join in theſe patriotic gifts; and by ſuch proceedings, the neceſſity of calling Parliament together, may be ultimately ſuperſeded. Such an alteration in the Conſtitution would, no doubt, be highly agreeable to the Alarmiſts.

To condemn opinions or inſtitutions for crimes which are committed by their real or pretended ſupporters, is both groſs fallacy, and flagrant injuſtice. The moſt clear principles, and the moſt ſalutary inſtitutions, are open to the attack of ſuch ſophiſtry. Suppoſe I were inclined to condemn the Britiſh Government in Church and

State

State, becaufe the riots and exceffes in Birmingham were in behalf of Church and King, would
any man of fenfe admit fuch events as full proof
that our Conftitution is fundamentally vicious ?
Yet, if I were to argue thus, I would only imitate Mr. Young. I might like him fay, " the
" *theory* of the Englifh religion is *peace*, but the
" *practice* is *riot* ;—the preachers of it, tell us
" to have good will towards our neighbours, yet
" they inftigate a mob to fire and plunder them.
" As for their kingly government it is ftill worfe.
" The people pay about twenty millions an-
" nually for *protection*, but at Birmingham the
" rabble are permitted to *burn* and *rob* for a
" week together ; and if it fo happens that
" they have deftroyed your whole property,
" and you have not a long purfe to go to law,
" you can have no redrefs :—If you are com-
" pletely ruined, you are completely excluded
" from juftice.—Such is the Britifh Govern-
" ment—Such is their glorious Conftitution !—
" But in preference to it give me the fimple
" defpotifm of Pruffia : At Spandau I may go
" to fleep in fafety :—at Birmingham I may ex-
" pect to be awakened amidft the flames of
" their hellifh Church and King Government."
In 1780, Mr. Young might ftill have condemned the proteftant religion—not more juftly—
but more fpecioufly. If he had written on the
6th of June, in that year, he might, with
the fame fairnefs with which he treats the
French, have faid, " The proteftant religion is
" a bayonet in your breaft, or a bullet in your
" bofom. In *theory* it teaches to do unto others
" as you would have them do unto you ; in
" *practice*,

" *practice*, it teaches fire, plunder, devasta-
" tion and bloodshed. The Protestants *entreat*
" redress *by petition*, and assault, wound, and
" threaten those who are to grant it : They
" pray for their enemies, and burn the houses
" and chapels of those who do not offend them :
" They declaim against drunkenness, and get
" so beastly drunk that they are consumed in
" the diabolical flames of their own kindling :
" they recommend honesty and mercy, yet
" empty the goals of felons and murderers
" whom they make their associates. Their
" *theory* instils *virtue*, but their practice ap-
" proves only of *vice*. They enlighten mankind
" by a general conflagration, and send souls to
" heaven by bludgeon and halter worship. Such
" is the protestant religion ! Give me paganism.
" Among the Tartars I may sit in safety under
" my own fig-tree. No sanguinary rapacious
" protestant, with the creed in one hand and a
" bludgeon in the other, will, in Thibet, de-
" mand money at my door and threaten to burn
" my house : THERE I WOULD BE IN SAFETY."
In this manner Mr. Young might condemn the
British government and religion, by blaming
them for the crimes committed in popular com-
motions, which the best of governments cannot
always prevent ; and from these as reasonably
draw conclusions unfavourable to them, as he
has done to the French Revolution, by applying
the crimes of individuals, or the temporary ac-
cidental confusion and mischief in France, to the
principles which actuate the mass of the people.
Our Alarmists have not only adopted the
French system of ruling by Clubs and Proscrip-
tions,

tions, but our own, and the other governments
in Europe, have difcovered much that is worthy of
imitation in the conduct of the unanointed executive Council. If the French affiliated Brabant by
the point of the fword, have not the allies bullied,
infulted, dragooned every neutral power ? Sweden, Denmark, Venice, Switzerland, Tufcany,
and Genoa, have had their independence violated. I dare not, indeed, fpeak in adequate
terms of the conduct purfued with the two latter
powers, whofe towns were threatened, with to-
lerable plainnefs, to be given up to military execution, if they did not declare War againft
France ! Is there no refemblance to be found
between the decree of the 19th of November,
and the conduct of our Generals and Admirals in
the Weft Indies and Toulon, where they have
offered affiftance and fraternity to all Frenchmen
who would renounce obedience to the mother
country, and accept of TRUE liberty, as modelled
by thofe who have juft beftowed it on Poland ?
The more the proceedings of the Belligerent
Powers are examined, the more it will be found
that France, which formerly fet the fafhions of
drefs, now fets the fafhions of government over
Europe ; and that the people of England in particular, have adopted their political conduct, with
more eagernefs, than ever they adopted the ftile
of a cap or a coat. The objects, indeed, are, different ; but the means taken to obtain or fecure
thofe objects, and which alone excite indignation, are the *very fame* :—Nay, the objects of
the French people, the eftablifhment of the Liberties, the Peace, and the Happinefs of Mankind,
are *good*, admitting the means to be bad : But
the

the objects of the Tory Jacobins, perpetuating the ignorance of the poor, arming the rich againſt the poor, the ſuppreſſion of the Freedom of Speech, and the Liberty of the Preſs, together with the juſtification of every vice of government, are *bad*, and their means are *equally bad* alſo.

It will be found, upon a candid examination of facts, that inſtead of the crimes which have diſgraced France, being the conſequence of her principles, they have been produced and wilfully inſtigated by the German Princes, in hopes of rendering odious thoſe very principles, as they dreaded that their eſtabliſhment in France, would ultimately deſtroy all tyrannical governments in Europe. The French were by no means the authors of the preſent War; on the contrary, they did all in their power to prevent it. It was the Courts of Auſtria and Berlin, and a few profligate emigrants, that provoked the rupture, which has now involved Europe in calamity. France confined herſelf within her own territories, and to her own domeſtic concerns, till ſhe was exaſperated by a combination againſt a government recently ſettled, approved by her inhabitants, and which promiſed to eſtabliſh a peaceable and laſting limited monarchy. Before the ferment of the firſt Revolution was allayed, and the dregs it had ſtirred up were ſunk to their proper obſcurity, the King, by his flight, proved his connection with the Princes who were plotting againſt the Conſtitution. Their hoſtile intrigues at foreign courts, became too manifeſt and provoking: they deſerted their country to procure its invaſion, and

left

left the field compleatly in poffeffion of the Republican party, whofe credit they ftrenghtened by their own treachery, and taught the people to believe the Nobles naturally and irrevocably their enemies; and to conclude, that fecurity for their Liberties was to be looked for, in their own interference and activity, rather than in the generofity or juftice of the higher ranks of fociety.

In September 1791, the manifefto from Monfieur and the Comte d'Artois to the King of France, announced their fuccefs in perfuading the Emperor and the King of Pruffia to hoftilty againft the new Conftitution : and thofe Monarchs gave it under their hands, figned at Pilnitz, that they required all the powers in Europe to affift them in this War, for " the Rights " of Sovereigns." The Princes alfo affirmed, that " the *other* European Courts had the fame " difpofitions with thofe of Vienna and Ber- " lin."—The fpeedy confirmation of what they had afferted, refpecting the Emperor and King of Pruffia, procured belief to the other paffages of their declaration, particularly as they remained uncontradicted by thofe to whom they made allufion; and it was neceffarily concluded, that all courts, and even that of St. James's, wifhed well to the concert of Princes. But how did France act in this fituation ?—Not like England laft winter. She did not precipitate herfelf into a War, although provoked by the moft flagrant violation of the Rights of Nations. She negociated;—for eight months fhe negociated:—till finding it impoffible to obtain

E   either

either fatisfaction or explanation for the Pilnitz Confpiracy, and that the Defpots of Germany were refolved on hoftilities, and were only delaying their commencement till fully prepared, fhe fubmitted to inevitable compulfion; and to fhew that fhe was not difmayed or terrified, fhe proclaimed War, which on her part was indifputably *defenfive by anticipation*. But notwithftanding hoftilities had been effectually declared by the Emperor and King of Pruffia, in Auguft 1791, and fhe did not refolve on a rupture till the April following, many perfons have had the effrontery and folly to affert, that France invited the War!

It is neither to their new principles, nor the natural cruelty of the French, that we are to afcribe the atrocities which have lately difgraced that nation. It might as well be faid, that the principles of the Proteftant Religion inculcate murder and rapine, becaufe fuch doctrines have lately been recommended from the pulpit; or that they produced the riots in 1780, as that the Principles of Liberty which have animated the French are the caufe of the crimes committed by fome of them. It is to the foreign combination againft the people, and the refiftance of the rich to the eftablifhment of their Liberties at home;—" It was that hateful out-
" rage on the rights and feelings of human
" nature, that wretched tiffue of pride, folly,
" and inhumanity; it was the Duke of Brunf-
" wick's Manifefto that firft fteeled the heart and
" maddened the brain of all France; which pro-
" voked thofe it had devoted, to practice all
" the

" the cruelties it had impotently threatened to
" inflict ; which sharpened the daggers of the
" affafins of the 2d of September, and whetted
" the axe fufpended over the unfortunate Mo-
" narch *." That infamous manifefto produced
the 10th of Auguft ; the treacherous furrender of Longwy provoked the horrid maffacres
of the 2d of September ; the inevitable War
with Britain and Holland hurried the unfortunate Louis to the fcaffold ; and the lofs
of Valenciennes proved the forerunner of the
trial, and the delivery of Toulon the fignal for
the execution of the Queen. The fuccefs of
the combined powers has invariably occafioned
the very reverfe of their profeffed object ; it
has always made more enthufiafts, united the
people more firmly, and removed the profpect
of re-eftablifhing Monarchy and Nobility to a
greater diftance. Since all Europe has joined
in the cry of erafing France from the lift of
nations, fhe has only retaliated by vowing the
deftruction of all Defpotic Governments. The
monfters of France have been begotten by the
monfters of Germany. The Duke of Brunfwick firft taught them profcription and murder ; their only choice was, whether they fhould
affaffinate, or be affaffinated ;—whether the
Duke of Brunfwick fhould prefide over another
St. Bartholomew, or Marat over a 2d of September.

We fhuddered at the news of the 2d of September, yet the *merciful* doctrines preached at
home, did not even excite a murmer. On the

Royal-Exchange the gallows was openly talked
of, as neceffary to extirpate the favourers of
Peace and Reform. If then, Englifhmen, by
a falfe alarm, could be guilty of wifhing for the
blood of perfons whofe only crime was difference of opinion, why fhould we condemn the
French, in whom every evil has been realized,
which here was only dreaded? If Englifhmen,
merely becaufe they were *told* they were in
danger, could endure fuch fanguinary language,
what might not have been their acceffes if fome
of the principal perfonages in the kingdom had
brought an hundred thoufand German robbers
to deftroy our valued Conftitution and eftablifh
Defpot ifm? If they had taken poffeffion of Yar-
mouth and Norwich, and threatened to give
up the city of London to " military execu-
" tion," what mad or wicked theorift might
not then have been liftened to? Mr. Young,
by warning Britain againft the example of
France, does in fact acknowledge, that in fimilar
circumftances, Englifhmen would be guilty of
fimilar enormities. If they would not, what
occafion is there for *warning* them? The inhabitants of France are not naturally more cruel
than thofe of this country, and were the enemies of that nation to ceafe to goad and exafperate her by their forces without, and their
treacherous and incendiary agents within, a fair
experiment might then be made on the undif-
turbed operation of French principles; it might
then be feen, that from the forced ftate in
which France is held by the confederacy of
Defpots, arife thofe crimes, and that impunity

to the Principles of Liberty. She might then, indeed, be an example, inftead of a warning to other nations: and nobody forefaw this more clearly than the Defpots themfelves; for when the Conftituent Affembly declared in favour of a pacific fyftem, againft conqueft or offenfive War, the whole fraternity on the continent were alarmed for their *trade*. Abolifh War! Abolifh the means of gratifying our ambition, of plundering our fubjects and perpetuating our tyranny!—Abolifh War!—Then we muft make War on you to prevent it. We will give a ftab to your pacific fyftem in its infancy, and by driving you into exceffes and horrors, convince the world that War, our Trade, is neceffary to the Happinefs of Society.

I cannot clofe the whole fubject of the War between France and the other continental powers better than by quoting what Lord Mornington faid in the Houfe of Commons, May 7, 1793, of the War in which England was engaged immediately after the revolution of 1688: his words may very juftly be applied to the prefent conteft, if inftead of Louis XIV, we read the Triumvirate, the Emprefs of Ruffia, the Emperor of Germany, and the King of Pruffia. "The Wars which " immediately followed the Revolution," (faid his Lordfhip,) "were abfolutely neceffary for the fe- " curity of that aufpicious fettlement. The re- " cent eftablifhment ·required protection, not " only againft the abdicated King, and thofe " who fupported his caufe at home, or who " had followed his fortunes abroad, but alfo " againft the pride and jealoufy of Louis XIV. " *He could not fuffer a LIMITED* MO- " NARCHY,

" NARCHY, *FOUNDED* ON THE LIBERTIES
" OF THE PEOPLE, to grow up and flourish *in*
" *peace* so near his Throne. He naturally fore-
" saw that such a Constitution must become a
" *continual reproach* to the tyranny of his go-
" vernment, and an insuparable obstacle to the
" progress of his ambition. It was, therefore,
" his policy to attempt the destruction of so for-
" midable a neighbour, by every means both of
" open force, and *secret machination.*"

Next to the crimes committed in France,
which have artfully been misrepresented as the
consequence of French Principles, the word
Equality has been falsely deemed to mean an
Agrarian Law : thus at once exciting horror,
and creating an alarm for property. Mr.
Reeves, in the publication already alluded to,
says, Equality means, that all men shall be
equally tall, equally wise, and equally strong,
as well as equally rich. This may be fit for Mr.
Reeves to say, but not for me to answer. I do
not doubt, however, of its having prejudiced the
vulgar, who would naturally believe the French
could be guilty of any thing, after being gravely
told, that they had roasted alive, basted with oil,
and eaten many of the Swifs and Nobles ; that
others they had made into pies and cried them
about the streets*; and that they " were drunk
" with the blood of their fellow citizens†."
Mr. Young makes a better use, for the Tory Jacobins, of the word Equality, by construing it
to mean an equalization of property ; " for,

---

* Vide the accounts of the transactions on the 10th of
August, published foon after that day in the Times.
† Vide Bishop of Gloucester's Sermon.

fays

fays he, " that all were equal in the eye of the
" law, was decreed by the Conftituent Affembly,
" and why call the year 1792 the fourth of
" Liberty and firft of Equality?"—All were not
equal in the eye of the law by the decree alluded
to, or even previous to the 10th of Auguft.
One whole branche of the Legiflature, the King,
was above the law ; and although I am ready to
admit that it was childifh to make the diftinction
in dating the year, yet, when I find fome reafon
for it, and none in fupport of Mr. Young's affertion, that " Property was glanced at," I think
it is proper to take that reafon, however fmall,
as the fair explanation, rather than put a conftruction upon the word which no circumftance
juftifies, merely becaufe Mr. Young afferts " it
" either meant that (equalization of property) or
" it meant nothing ;" becaufe he acknowledges
that he does not underftand it, fhall we receive
as fact what he wifhes, and for that reafon fuppofes
it to mean ? The King was not amenable to the
law, therefore, ftrictly fpeaking, there could
not be perfect Equality. On the queftion of
Royal inviolability I give no opinion.

It is not neceffary to fay much in anfwer to
the common mifreprefentations of Equality. No
candid intelligent man, either in France or England, could ever underftand it to mean otherwife
than an equality of refponfibility to, and protection from the law ; and of this, no nation
in the world was ever more in want than France
previous to the Revolution in 1789. The expence of law-fuits conftitutes the only grievous
inequality in England. It is impoffible for a
poor man, with his own means, to obtain redrefs

dreſs by a courſe of law in this country* ; and, even a prudent man of ſmall fortune, will ſubmit to an injury, rather than riſk his ruin by entering into a conteſt with a wealthy litigious opponent. Many an inſtance of injuſtice, accompliſhed and maintained by riches, may be found in the Memoirs of a northern Nobleman, who has long been the tyrannical oppreſſor and plunderer of the poor, and middling claſſes, in that part of the kingdom, which is curſed with his reſidence.

It is to what *he calls* the French Equality, that Mr. Young aſcribes a ſyſtem of rapine and plunder, which he affirms exiſts in that nation. He concludes a number of his paragraphs with " This " is Equality ! Rob the farmer ! Plunder the " landholder and divide his land ! Equality of " property ! An Agrarian Law, &c. According " to French principles, the firſt beggar I meet, " may ſabre in one hand, rights of man in the " other, demand a ſhare of what is my own, " my property, my land at Bradfield." And in this manner he proceeds, arguing, as if an equalization of property had actually taken place

---

* It has been ſaid, even from the Bench, that our laws are equally open to the poor as to the rich, becauſe if a man cannot fee an advocate, the court will provide him one gratis. This, like ſome other parts of our Conſtitution, was excellently intended, and, no doubt, at its firſt inſtitution, proved highly beneficial ; but like many more inſtitutions, Time, that great Innovator, has reduced it to a mere ſhadow. The fee of a counſellor is the leaſt expence of a law-ſuit ; nor can his ſervices be of much utility to his client, if he is unacquainted with his caſe, till after the trial has begun ; and even then moſt probably, he cannot, from the ignorance of the poor man, rightly underſtand it.

in

in France ; or at leaſt as if ſome notorious
and undeniable proofs had occurred of robbing a
man of his eſtate, and dividing it among
" Beggars."

The only inſtance, however, which he attempts to produce in ſupport of this, (and if
more or better could have been found, he has
ſhewn ſo much induſtry in collecting the crimes
of the French, that he certainly would have
brought them forward), is, an unauthenticated
one in the Clermontcſe, where an eſtate was
ſeized from the proprietor, who was, no doubt,
an emigrant, as Mr. Young owns, " *he lived at*
" *a diſtance*"; and where ſome of the tenants
who talked of retaining their farms, becauſe thoſe
who laboured ſhould not pay morcy to thoſe
who did nothing, were, moſt probably for ſo
doing, obliged to quit, and the property was
ultimately held in truſt for the nation. But
this ſtory reſts wholly on Mr. Young's bare aſſertion, who has not quoted even the authority
of a newſpaper, which he invariably does on all
other occaſions ; and whether it is true or falſe
is immaterial, becauſe it proves nothing like a
general principle of equalizing property. In
the preſent ſtate of France, no doubt, many
atrocious robberies are committed, which would
be puniſhed or prevented, if the German Plun-
ders would allow the government to act, as it
certainly would do in tranquillity, with rigo-
rous juſtice. In any country in a ſimilar
ſituation, ſimilar crimes would be perpetrated ;
and at all events they can have no connection
with the Principles of Liberty and Equality.—Becauſe the northern Lord to whom I have juſt

F   alluded,

alluded, has, by dint of wealth, plundered every
man in his neighbourhood who had any thing
to lose, are we therefore to conclude, that it is
a principle in the English law that the rich may
rob the poor?

Instances, and strong authenticated instances,
to confute Mr. Young's affertions respecting an
Agrarian Law are on record, and cannot be denied. About the very time he was writing his
book, when all the horrors and anarchy, on
which he continually rings, were reigning, the
Duc de Penthievre died, where he had long
lived, in the heart of France. He posseffed
immense wealth both in land and moveables;
was the father-in-law, but enemy, of the duke
of Orleans, from whom his daughter had been
long feparated; and he was even fuppofed to
be unfriendly to the Revolution. He was an
amiable and beloved man; he never interfered in
politics; amidst every violence and change, his
property and perfon were untouched, and many
a poor man's tear bedewed his grave. No beggar, fabre in hand, demanded a fhare of his property, either during his life or at his death.
Here then, is a direct notorious proof, that
the property even of an *eminent nobleman* is
not violated, while he does not oppofe the exift-
ing government; and undoubtedly every man,
who conducted himfelf like the Duc de Penthievre, was equally unmolefted. It was in-
cumbent on Mr. Young to give proofs, that men
had been plundered in France for no other offence than that of being rich, before he pretended to be ruled by " Events" which never
took place, to be guided by " Experiments" which
never

never were made, and to follow a " Practice"
which never had any exiftence, but in his own
credulity or mifreprefentation.

It is, however, highly probable, that the
French Government, fince it has no ally to pay
the expences of defending its own territories,
and is fo hemmed in and infulated by its enemies
that it cannot borrow, will be obliged to have
recourfe to fome extraordinary meafures which
may be no further juftifiable than on the hack-
neyed plea fo often ufed by its prefent invaders,
" State neceffity !". But whatever may be done
in the prefent ferment and alarm, when France,
inftead of thinking of ftrict juftice, has to ftruggle for her very exiftence as an independent nation, will be no proof that fhe would not have
been, if permitted to remain in peace, and yet
will be, when peace is reftored to her, as equitable as her neighbours. Should the Government feize a part of the effects of the wealthy
for the defence of the country, that will give
no more reafon to expect an Agrarian Law,
than raifing twenty Millions annually in Britain gives reafon to expect it will be afterwards
equally divided among the people. Mr. Young
well knew that there is a decree, making it
death to propofe an Agrarian Law, which has
long been and ftill is in *full force*, but it did
not fuit his impofture to notice it. Whatever
may be done in France, ought not to be afcribed
to the principles or rapacity of the people, but
to the unparalleled exigency of the times. To
fupport her Liberties, indeed, the rich will probably be forced to pay what the poor cannot ;
and to this Mr. Young will not object, as he ad-

mits,

mits, p. 123, " In taxation, fpeaking at large of
" a nation, to quantum *paid* is not fo? much
" the object to regard as the quantum *left* after
" the taxes are paid."

But while I deny that any thing like proof
can be produced of an Agrarian Law having
taken place in France, I own that an extenfive
and awful confifcation has been made! What
has been the occafion of it?—Not a defire to
plunder and divide Property. It has been the
folly, weaknefs, or wickednefs of thofe who,
have fuffered; fome of whom openly declared
againft the Conftitution, and excited the invafion
of the country; others had made their hoftile
difpofition too manifeft to be fafe at home,
therefore they fled; and the ruling power in
Paris, finding the Emigrants were drawing out
the wealth of the nation in order to make
War upon it, feized the eftates of all thofe
who did not appear within a limited time. Such
as were friendly to the Conftitution could have
no objection to appear at their poft, becaufe they
were in no danger; and thofe whofe hoftility
was afcertained by their abfence, deferved to lofe
their property; deferved to lofe it on the principles of the Britifh Government, which feized
the property of the Rebels in 1715 and 1745, as
greedily as the French Convention; and whofe
tremendous confifcations in Ireland during
the Commonwealth, (afterwards *confirmed* by
Charles the IId.) and in the reign of William
the IIId. furpafs all that has hitherto been done
in France; where, if mens' minds were reftored
to tranquillity by peace, and their country fecure, the vengeance of enthufiaftic fury would
aſſuage,

affuage, and probably moft of the fugitives would be reinftated in their ancient inheritance and fequeftered property.

Befides general pictures of French exceffes, and mifconftructions of the word Equality, Mr. Young continually maintains " Experiment," " Practice," and " Events," to be the only wife rule of conduct ; and, alluding to the experiment of perfonal Reprefentation being attempted in France, for I deny that it has been made, he fays, (p. 56) " The thing is tried ; that method " of drilling has been experimented and found " good for nothing ; the crop did not anfwer." But if two neighbouring farmers were to try a new theory, and the children of the one were to drive beafts into their father's field, to nip the firft fhoots, to trample and deftroy the corn as it fprung from the earth, and his crop did not anfwer ; while that of the other farmer, whofe field was unmolefted yielded an abundance beyond all expectation, could it, in fuch a cafe, be juftly faid the experiment failed ? The farmers are France and America ; the children are the Emigrants, and the beafts are the Germans. The experiment of Reprefentative Government has not been allowed a fufficient period to be made in France ; for even if the field had been unmolefted, the corn has not had time to vegetate, much lefs to fill and ripen. In America, the field was as much difturbed, during feveral years, as in France, and accordingly produced nothing but weeds. But the theory was known to be good ; it was perfifted in, and when fair play was given to the foil, the crop was aftonifhing. So it will be with France, which at prefent,

fent, in refpect to Government, may be called
a neglected or barren foil; for there is a more
important duty than cultivating the field; the
natives muft defend it, or they will have no
field to cultivate.

But fays Mr. Young, the experiment has not
finally been tried in America. No! Eighteen
years is no experiment in America, though fix
months is a compleat experiment in France!
" But America has fuch a plenty of land that
" fhe has no poor; it is not her Reprefentative
" Government that prevents there being any
" poor, but it is the plenty of land. When fhe
" has a numerous indigent poor, her Govern-
" ment will tumble to pieces; the mob will
" not poffefs the Sovereign authority and re-
" main hungry," continues he. I deny that it
is her extent of territory alone that prevents her
being burthened with a numerous indigent poor.
Look at almoft every country in Europe, not
even excepting Britain and Ireland, and acres
enough will be found to make thofe who are
now poor as comfortable as the American farmer,—I do not mean by an Agrarian Law; I
mean by properly employing the wafte unoccu-
pied lands. Look at the vaft tracts in Germany,
Hungary, and eaftward towards the Black Sea,
whofe native fertility is doomed, by a barbarous
policy, to feed wild beafts inftead of man; where
whole countries, that would render millions happy, are made a blank on the earth by the reftlefs
defolating ambition of bad Governments, or
perhaps, to afford amufement to one Tyrant. In
Spain, the example of pride and indolence in
the Nobles, has occafioned fuch a contempt and

neglect

neglect of Agriculture in the lower classes of society who ought to attend to it, that population has decreased, and the nation become insignificant in the scale of Europe compared with what it formerly was. The natives live chiefly on the natural productions of the soil, one of the richest in the world, which if properly cultivated, might maintain in greater plenty than at present, six times the number of inhabitants. " The extent of ground is of so little value " *without labour.*" says Mr. Locke, " That I " have heard it affirmed, that in Spain itself, a " man may be permitted to plough, sow, and " reap, without being disturbed, upon land he " has no other title to, but only his making use " of it. But on the contrary, the inhabitants " think themselves beholden to him, who by " his industry, on neglected, and consequently " waste land, has increased the stock of corn " which they wanted."

It would be superfluous to point out similar evils in the other European nations. With regard to France, under its old Government, which is more immediately interesting to the question, I desire no more to be said in support of my opinion than what is contained in Mr. Young's own travels. It will there be found that it was not the great population of that country which occasioned the poverty and wretchedness of the multitude, but that it was the monopoly of the land by a dissolute tyrannical Nobility and Priesthood, whose possession operated effectually to diminish the produce of the soil, and to blast with sterility those plains upon which nature had lavished her richest bounty. The arbitrary and

enormous

enormous exactions of the Agents of the Government aggravated the diftrefs of the people, among whom lay acres fufficient to make them as happy as the American farmer. It is the miferable policy of Governments that makes fo many poor in Europe ; and what is the caufe of the people being in general much more happy and wealthy in Britain than in Germany ? Not " Corruption," as thofe who fatten by it would make us believe. It is becaufe we have much lefs of the old fyftem remaining ; becaufe men are more enlightened, induftry encouraged, and property better fecured ; becaufe our anceftors have wrung from the hand of power and bequeathed us more Liberty than is enjoyed under any other Monarchy ; nor can thefe bleffings be preferved but by that fpirit which acquired them.

But it may ftill be faid, America has more acres in proportion to the number of its inhabitants than any country in Europe. I fhall not difpute this, becaufe it is of no importance to my argument. I contend that there is a fufficiency of acres in Europe to make every man as comfortable as the American farmer, and that the extent of territory in the New World does not alone prevent indigence among the people. It is the fpirit of their Government, which encourages not only Agriculture, but Manufactures and Commerce, and difcourages War. Look at Holland ! The number of poor in Holland is perhaps as fmall as in America, and certainly much lefs than in any other European nation ; yet Holland contains very few, and thofe very unproductive acres. It is the moft populous fpot in Europe, and, leaving her poor out of the

the calculation, is twenty times more populous than America. How happens it then, that with fo few acres, and fo many inhabitants, the multitude of the Dutch are even more wealthy and comfortable than the American farmer? Becaufe, though her Government is very defective, its fpirit encourages induftry, commerce, and œconomy, and always avoids War, unlefs forced into it by fome infidious friend. It is the encouragement given to Manufactures and Commerce, as well as Agriculture, rather than the quantity of acres, that makes a people rich and happy. Dr. Smith, and all profound political œconomifts, affirm, every Artizan, Manufacturer, and Labourer, to be as valuable to a ftate as acres of land, and that the greater the population, if Government animates and properly directs induftry, the greater will be the riches of the country, and enjoyments of the people. It is wicked pernicious policy alone that proves, and fuch fhallow, or deluded politicians, as Mr. Young, who affert, that population is the caufe of poverty, and that the richeft man who ever lived, was Adam, becaufe he alone exifted, and was landed proprietor of the whole world.

The view Mr. Young takes of America, with regard to paying taxes, is equally unfair with his " *Speculations*" refpecting her Reprefentative Government. · He fays, that a Farmer in the Back Settlements may have plenty of beef, mutton, corn, wool ; he may be rendered quite eafy and happy, by a fuperfluity of the neceffaries of life, but for want of Circulation, of Commerce, he will not be

able

able to pay a fingle Tax ; and therefore when America is involved in a War, and called upon for Taxes, fhe will be ruined. *When fhe is involved in a War!*—But her Government will not every few years alarm, miflead, and madden the people into one, becaufe the Government emanates *from* the people, *for* the people's benefit, and is not directed by an ambitious or felfifh junto:—And it is unjuft in Mr. Young to eftimate the power of America to pay taxes by the farmers in the *back fettlements*. He owns, that the duties laid on the diftilleries in Scotland, do not pay for the collection, and fays, that the expence of collecting taxes from the farmers in the back fettlements, would alfo be more than could pay for their collection : would it not, therefore, be as fair to conclude, that Britain cannot pay taxes, becaufe the Scotch diftilleries cannot, as to conclude that America cannot pay taxes, becaufe the farmers in the back fettlements cannot ? A little confideration will fhew us, that America, like Britain, has circulation and commerce; like London, Liverpool, Briftol and Glafgow ; it has Philadelphia, Bofton, Charles-Town and New York:—thefe laft, indeed, may not be fo extenfive and rich as the former; but, two hundred years ago, London was not a fourth part fo wealthy and populous as at prefent ; and is not America daily increafing her commerce, manufactures and population ?

Thus, I think, Mr. Young's " Experiment," " Practice," and " Events," as applied to America, are wholly illufory. Reprefentative Government has ftood in Peace, and flourifhed in
America

America twelve years. " Yes," fays he, " but " it will be deftroyed when fhe has a numerous " indigent poor."—*When* fhe has an indigent poor ;—but, that will never be, more than at prefent, while the fpirit of her government prefers, as it now does, Peace and Induftry, to War and Corruption. He would condemn Reprefenta-tive Government from the " experiment" of a few months attempted in France,.in the midft of the moft dreadful warefare ever known ; but in America, " Experiment," " Practice," and " Events," which he fays, ought to be the only rule of conduct, he would fet wholly afide, becaufe *there* he finds they are ftrongly againft him, and that he cannot hold up a *terrifying example* of what *has* happened. He would condemn RceprefentativeGovernment in America on " fpeculation" and " theory," which he a thoufand times fays, ought never to guide us ; he would condemn it by predicting, what *will* happen, not by fhewing, what *has* happened. On the contrary, in France, the experiment of a few months, is to be our only guide, and theory and fpeculation we are totally to difregard ; in America theory and fpeculation are to be our only guide, and the experiment of twelve years we are totally to difregard ! ! !—How Mr. Young can reconcile thefe palpable contradictions, or how any man, of the leaft underftanding, could be duped by them, is beyond my comprehenfion. It may juftly be afked of him, in his own words, "what " inducement have we, therefore, to liften to " *your fpeculations*, that condemn what all" *America* " feels to be good ?"—(p. 85)

Hence

Hence, I think, I may fairly affirm, that in America, there is a noble instance of the blessings flowing from Representative Government. The Revolution there, was at first, as much execrated as that of France now is. In 1777 the same invectives may be found in the proclamation of general Burgoyne, that in 1792 were brought forth in the manifesto of the Duke of Brunswick : " He appealed to the suffering " thousands in the provinces, whether the pre- " sent unnatural rebellion has not been made " a foundation for the compleatest system of " Tyranny that ever God, in his displeasure, suffer- " ed for a time to be exercised over a froward " and stubborn generation. Arbitrary im- " prisonment, confiscation of property, perse- " cution, and torture, unprecedented in the " inquisitions of the Romish Church, are among " the palpable enormities which verify the " affirmative. These are inflicted by assemblies " and committees, who dare to profess them- " selves Friends to Liberty, upon the most quiet " subjects, without distinction of age or sex, for " the sole crime, often for the sole suspicion, " of having adhered in principle to the govern- " ment under which they were born, and to " which by every tie, divine and human, they " owe allegiance. To consummate these shock- " ing proceedings, the profanation of religion " is added to the most profligate prostitution of " common sense ; the consciences of men are " set at nought, and multitudes are compelled, " not only to bear arms, but also to swear sub- " jection, to an usurpation they abhor." Here is an epitome of all that has been said and written

against

againſt the French, and why ſhould we not ſuppoſe, that were they left to themſelves, they would
ſoon become as orderly, peaceable, flouriſhing,
and happy, as the Americans? France has the
natural ſources to make merchants like Holland,
manufacturers like Britain, and farmers like
America. And, I believe, the concert of Princes,
was firſt formed againſt her from a knowledge
that a few years Peace, would make her people
ſo wealthy, comfortable and happy, that inſtead
of being a terrible warning, ſhe would be a
ſeducing example to ſurrounding nations; that
the Deſpots of the continent ſaw ariſing from
the ruins of French Deſpotiſm, the fabrick of
human felicity, and conſequently the harbinger
of their deſtruction, and therefore they wiſhed
to ſtrangle it. At preſent, France is neither an
example, nor a warning; ſhe is in a ſtate of
madneſs, occaſioned,—not by perſonal Repreſentation, as Mr. Young continually aſſerts, for
he might as juſtly ſay, the ſhocking yellow fever
which has juſt broke out in Philadelphia, is the
effect of Repreſentative Government in America;
but, by the infamous invaſion of the Duke of
Brunſwick, and the treachery and alarm raging
in her boſom. I therefore affirm, that an ad-
mirable proof of the utility of Repreſentative
Government exiſts in America, and deny, that the
experiment of perſonal Repreſentation has been
even *tried*, much leſs " *compleatly made*" in
France, or that ſhe ſhould either encourage, or
deter us, from ſalutary Reform. France would
not Reform till it was too late:—may her example, in that reſpect, be, indeed, a warning to
Britain.

Having

Having shewn Mr. Young's fallacy in condemning Representative Government in France upon " experiment," and in America upon " speculation ;" and having established the experiment of America in my own favour, I shall now notice a few of the most mischievous falsehoods, and absurd contradictions, which he has advanced against Reform, but without stopping to examine every groundless assertion, and trivial argument. I shall take his leading reasons only; the most unfounded and atrocious of which is, where he maintains, " that the " example of the King of France should deter " all other Kings from listening to the com- " plaints of their subjects; he wished to Re- " form; he was the first Monarch who was " desirous of making his people happy, and *he* " *died for it on a scaffold*."—It is difficult to say, whether the falsehood contained, or the Tyranny recommended in this passage, should most provoke indignation ! The falsehood is so well known, that none but Mr. Young could have had effrontery sufficient to have asserted it; and the sentiment conveyed is more dangerous to the liberties and happiness of Britain, than all the works of Mr. Paine and the " Jacobin Societies." The King of France *did not* die on the scaffold because he listened to Reform ;—it was because he would *not* listen to it till it was *too late* ;— till he was *compelled*. Instead of listening to Reform, he listened to evil Counsellors ; he suffered himself to be guided by persons who lavished the puplic money on useless placemen and pensioners ; on favourites, parasites, and the most pernicious of all traitors, *regular-bred courtiers,*

courtiers, who, to supply their prodigality, loaded the people with Taxes ; and finding, at last, their profligate extravagance was on the eve of ruining them, agreed to Reform as their only chance of salvation. They first threatened the Parliament of Paris if it did not comply with their demands ;—did that look like a wish to Reform ? They next tried the shift of the Notables ; but like their conduct with the Parliament of Paris, it also proved ineffectual; and after many other vain expedients, they were at last forced to call the States General, in hopes that they would cover their past iniquities, and grant a new lease for plundering the Nation. In that hope, however, they were also mistaken. The people had long smarted with patience under the Tyranny of the Nobles, and extortions of the Government ; they were now menaced with large additional burthens, and the general danger created a general alarm. The spirit of the country was roused by the infamy of the Administration ; and finding the Court only stooped to concession, to hide its mismanagement, and insure its future robberies, rather than from a sincere wish to Reform and make them happy, the people resolved no more to trust to those who had invariably betrayed them ; the cause of the people triumphed ; the rapacious views of the courtiers were defeated, and the old Despotism stripped of its riches and grandeur, was too feeble, ugly, and corrupt, to protract its existence. Such was the commencement of the French Revolution. It was not the desire of the King to *Reform*, but the desire of his Ministers to *rob*, that brought him to the scaffold.

All

All the calamities with which France is afflicted, are aſcribed, by Mr. Young, to perſonal Repreſentation, and the ſame are predicted in England, if any Reform of the Houſe of Commons takes place. Accordingly, a number of his paragraphs conclude with " Such is the " monſter, perſonal Repreſentation :"—" Such " the reſult of that Conſtitution founded on " perſonal Repreſentation :"—" Power, in the " hands of the people, by means of perſonal " Repreſentation, has ruined France :" &c.— And thus he proceeds, aſſerting, that all the misfortunes in which that nation is involved, are the conſequence of the will of the people, being ſupreme. In anſwer to this, which is the *ſole object* of his book, I ſhall ſhew, firſt, that he contradicts himſelf, and that the very reverſe of what he ſo often repeats and labours to eſtabliſh, may be proved by his own words ; and ſecondly, that the experiment of Perſonal Repreſentation has not been attempted in France, till *after* the downfal of Monarchy, and, therefore, it *cannot* juſtly be blamed with producing that event, or the tranſactions which preceded.

P. 91, he ſays, that "if the Houſe of Commons " were ſuch Repreſentatives" (as in France) " they *would be guided* by the folly, madneſs, " and paſſions of the people." Seven pages further on, however, he thinks quite differently, and that " Repreſentation deſtroys itſelf, and " generates with infallible certainty an oligarchy " of mobbiſh demagogues, till of all other " voices, that *leaſt heard* is the *real* will of the " people."—Again (p. 106.) he maintains, that " a Par-

" a Parliament conftituted on perfonal Repre-
" fentation *can* act *no otherwife* than by the
" *immediate impulfe* of the people." But this,
he ten pages before, affirms to be wrong ; he
thinks, " a word however, might be faid on
" the point of perfonal Reprefentation, rendering
" the will of the people *fupreme*. The *futility*
" of the idea is demonftrated, in the Affemblies *fo*
" *chofen*, in France ; their firft merit, on Jaco-
" bin principles, is, that of fpeaking the fove-
" reign will of the people, by which expreffion,
" is always underftood the Majority : But, fo
" truly abominable, is this fyftem of Govern-
" ment, that there has not been a fingle in-
" ftance of great and marked importance, in
" which the *Minority*, and commonly, a *very*
" *fmall Minority*, has not; by means of terror,
" carried all before them."—Thus, he at one
time affirms, that the will of the people " *would*
" guide" perfonal Reprefentatives, and yet,
that among fuch Reprefentatives, the will of the
people would be " *leaft* heard."—That perfonal
Reprefentatives " can act *no otherwife*, than
" by the *immediate impulfe* of the peo-
" ple," and that every inftance, of great and
marked importance, in France, has, by means of
terror, been carried by the will of a *very fmall
Minority !*—Thefe contradictions are fatal to
all Mr. Young has written, and are a juft
illuftration of the whole *fairnefs* of his book.

His inconfiftency, in condemning perfonal
Reprefentation, might be farther expofed, but
it will fhorten the fubject much to fhew, that the
cafe of France is wholly inapplicable, and is
neither an example, nor a warning to Britain,

H   with

with regard to Reform, particularly as applied
by Mr. Young. The Affembly which framed
the Conftitution of 1789, was not elected by
univerfal fuffrage, nor agreeably to any new fyftem
of Reform. It was elected according to the rules,
and under the direction of the French Monarchy;
according to a fyftem eftablifhed for centuries.
Reform, or Perfonal Reprefentation, therefore, are no more to be blamed for what it
did, that was blameable, than they would be,
if our Houfe of Commons were to exterminate
every Friend to Liberty, and eftablifh Defpotifm
in England. The Conftituent Affembly was
elected according to the old " Mild benignant"
French Monarchy; according to a fyftem,
" The work of the wifdom of ages; yet it
was that very Affembly which Mr. Burke reprobated, and Mr. Paine applauded, and againft
which the Convention at Pilnitz, the foundation of the prefent War, was formed. It
was the authority of that Affembly which the
Britifh Government would never fully recog-
nize: It is that Affembly which Lords Auckland
and Hood, and even Mr. Young defcribe, when
they mention " The mifcreants, who, for *four*
" *years*, diftracted France." Then why fhould
the tranfactions of that Affembly deter us from
Reform, fince it was elected according to the
Conftitution of the Old French Defpotifm? It
fhould rather be an argument *in favour* of Reform, becaufe, according to Mr. Young and
others, it produced much mifchief, and it was
an old UNREFORMED Affembly.

The fecond Affembly was not elected by univerfal fuffrage, and of the third Affembly, it is

uncandid

uncandid to speak, for, during its existence, it has rather been the Council of War of a great Army, than the seat of Peace and temperate Legislation : and it might as reasonably be concluded, that the British Government is despotic, because, for the sake of discipline, Despotism must be exercised in the British Camp under the Duke of York, as to conclude that Anarchy, Confusion, and Despotism, are the certain accompanyments of French Principles of Liberty, and will always be exercised by the Convention.

Having now shewn the inconsistency with which Mr. Young would build up his " Warn-" ing" against personal Representation, by which he always means Reform, having also shewn that the French Revolution, and the calamities afflicting that country, did not arise *from* personal Representation, which has neither had time nor opportunity to operate there ; and that therefore the transactions in France have no analogy whatever with Parliamentary Reform in this kingdom ; I shall, for a while quit French affairs to examine what Mr. Young says of the British Constitution, and particularly of the House of Commons.

" But" (says he, p. 92) " the House of " Commons are corrupted and bribed. And " if the nature of such an Assembly demands " to be corrupted in order to pursue the public " good, who but a Visionary can wish to re-" move Corruption ?" Must not an Assembly, constituted for the public good, be of a most *detestable nature*, if it demands to be bribed in order to discharge its duty ? Again, " In-" fluence, or, as Reformers call it, Corrup-

" tion,

" tion, is the oil which makes the machine of
" Government go well." And p. 171. " Ex-
" TRAVAGANT COURTS, SELFISH MINISTERS,
" and CORRUPT MAJORITIES, are fo
" intimately interwoven with our practical
" Freedom, that it would require better poli-
" tical Anatomifts, than our modern Reformers,
" to fhew, on fact, that we did not owe our
" Liberty to the identical Evils which they
" want to expunge." Could the whole National Convention more grofsly Libel the King,
the Minifters, and the Parliament, of this Country ? Surely, Mr. Young muft have known, that
he was writing the moft bitter and dangerous
fatire on our Government, when he faid, that
extravagant Courts, Selfifh Minifters, and *Corrupt Majorities*, were intimately interwoven
with our Freedom, and yet affert, that this is
" that glorious Conftitution which is the inhe-
" ritance and pride of Britons!" I appeal to
every candid man whether the friends to the
Liberty of the Prefs, or Mr. Young, feem moft
difpofed " *to publifh the Corruptions of the*
" *Conftitution*, in other words, *to write it*
" *down*. (vide p. 163.)

In anfwering this frontlefs avowal of Corruption, I muft fuppofe the Houfe of Commons compofed of either good or bad men. If they are
the former, and are fuffered to exercife their
own difcretion, they will purfue the public welfare without Corruption, unlefs it is prefumed
they cannot fee it without the explanation of a
Bank Note. If they are bad men, they will be
bribed to do evil as readily as good. It may indeed be faid, that influence is often neceffary to
                                          make

make a virtuous man do a virtuous action; and
that is true: For, although a virtuous man will
act virtuously when he does act, yet there are
occasions where he is not called upon by duty
to act at all: But on the contrary, a Member
of Parliament *is in duty bound* to act, to take
either one side or another; and if he is a good
man, he will act for the best, according to his
conscience, without either Corruption or Influence. It is an absurdity to say, that a man can
be *corrupted* to act *virtuously*; because if he
acts from a corrupt motive, he is equally vicious
whether the action be good or bad, and will not
much consider whether he does right or wrong,
while his primary object, Corruption, is obtained. The same reasoning will apply to the
assertion, that " a man may be bribed to act
" wisely;" unless the man is supposed to be a
fool. Nor do I understand how a man can be
influenced to do good, for, in a case, like that
of the House of Commons, where he *must* act,
he must be predisposed to do evil before he can
be influenced to do good, and consequently he
must be a bad man: And all the arguments in
favour of Influence, do, in fact, go to prove, that
the Members of the House of Commons, are
traitors to the State, who would ruin it if they
were not influenced to the contrary. As Mr.
Young has quoted Dr. Johnson's definition of
" *Principle*," I will take what the same author
says of " *To Influence*," which is, " to modify
" to *any* purpose." If Influence, therefore,
can modify to *any* purpose, (which it certainly
can, or else it is no longer Influence;) and if our
House of Commons is, as Mr. Young asserts,
directed

directed by Influence, then it is the mere crea-
ture of the Executive Power which poffeffes
that Influence ; and it is a miftake to fuppofe
that it makes laws, or does either a good or a
bad action : It is the King's Minifters, who,
holding the Influence that directs it, ought to
bv wholly blamed or thanked for whatever it
does; becaufe, they may *influence* it to the moft
wicked, as well as the moft wife meafures ; and
it would be better, if the " Monfter," as Mr.
Young calls it, were annihilated, for then
morality would not be wounded ; Government
would be carried on at a lefs expence, and with
more eafe and vigour ; and Minifters would be
much more refponfible for their conduct.

But while I maintain influencing or corrupting the Reprefentative Body by the Executive
Power (which in this country, is not, I hope,
the cafe, notwithftanding Mr. Young's affertions) to be the moft pernicious of all Policy,
and the blackeft of all Treafon; I am far from
maintaining that Members of Parliament fhould
not be rewarded for their trouble : On the
contrary, I think they fhould be openly paid a
regular, ample ftipend ; but it fhould be paid
merely as a reward for their trouble, and not as
an equivalent for their honefty ; they fhould
publicly receive a certain fum, and not fecretly
take a recompence which they are afhamed to
own. If it were the cuftom, that our Ambaffa-
dors fhould receive no pay from our own Government, and were permitted to take as much
as they could procure from the Courts to which
they were fent, can it be doubted that they
would facrifice our intereft for that of thofe
from

from whom they expected a reward? In like manner, it would be folly to suppose that a Parliament penfioned by the Government, would guard the interefts of the people against the Government.

I am at a lofs to conceive how the Corruption of the Houfe of Commons can be deemed the *caufe* of our profperity and happinefs, nor have I ever feen any fact produced tending to prove it. It is afferted, that we are profperous and happy, and admitted that Corruption does exift; and, therefore, it is inferred, becaufe Corruption is a part of the fyftem which makes us profperous and happy, *that it is the caufe of* our profperity and happinefs. But nothing can, I think, be more falfe than this inference : It might as juftly be inferred, that a mixture of weeds, among the corn, is the caufe of a plentiful harveft; that the drofs mixed in the ore with gold and filver, is the caufe of their value; or that infects and locufts are the caufe of a luxuriant herbage. Would a tradefman afcribe his accumulation of wealth to the pilfering of his till by his fhopman? Corruption may, indeed, be a *part* of a fyftem, good upon the whole; but it is a *bad* part, and ought to be removed. It is not to Corruption we are to attribute our profperity and happinefs, but to the fpirit and induftry of the people.

I fhall now confider the queftion of Reform under four diftinct heads, and in doing fo, fhall aim at brevity, and endeavour, as much as poffible, to avoid a repetition of the many arguments which have been advanced in favour of it. I fhall examine

The neceſſity of Reforming the Houſe of Commons.

How far a Reform can be made in conformity with the ſpirit, principles, and practice, of the Conſtitution.

The time for making a Reform.

What the nature of the Reform ſhould be.

Of the NECESSITY of Reform, the removal of Mr. Young's *ſalutary* Corruption, is, the ſtrongeſt proof; for Corruption, if practiſed, muſt deſtroy the whole original deſign of the Houſe of Commons, which, even according to to himſelf, was to *adviſe* the King in important matters of State. Now, as advice does not mean aſſent, but muſt often be contrary to the opinions and wiſhes of thoſe who aſk it; and as it is well known, that the King cannot act contrary to the advice of his Commons, it is, in fact, a controul, a command. If, therefore, Corruption or Influence makes them adviſe the King to whatever he recommends, their controul is loſt, and the King is as abſolute as the Empreſs of Ruſſia, ſo long as he poſſeſſes the means to *influence*. " If " Courts *can* be perfidious, you are to ſuppoſe " they *will* be ſo; and if you have not ſo pro- " vided as to turn that perfidy to the benefit of " the people, you confeſs at once, that your " Conſtitution is viſionary," ſays Mr. Young (p. 68,)—If ſo, what proviſion is there againſt the perfidy of the Britiſh Court, while it dictates to Parliament by its Influence? With Corruption and Influence, then, our Houſe of Commons muſt be an illuſion, and a dangerous illuſion; becauſe, it is not reſponſible, and takes off all reſponſibility from the Crown. It muſt

give

give a falfe appearance of fanction, and bind the
people to whatever the Crown propofes, and
pays for with the money taken from the people :
It would be as if a Court of Law were to
grant to a highwayman a licenfe to rob, upon
condition that he paid into it a part of his plunder. With Corruption, our Houfe of Commons, would be both *ufelefs* and EXPEN-
SIVE ; but I am far, very far from believing, it
is governed by Corruption, and I think it my
duty here to remark, on the fhamelefs effrontery
of Mr. Reeves, in thanking Mr. Young for
afferting, that we are governed by extravagant Courts, felfifh Minifters, and corrupt Majorities (vide p. 171.) and alfo, to exprefs my
furprife, that the Houfe of Commons has not
addreffed the King, or that the Attorney General
has not been commanded to profecute in his
book, that which in Mr Horne Took's petition,
refpecting the Weftminfter Election, was, by
all parties, deemed the moft grofs and fcandalous libel that ever came before them.

Mr. Young afferts, and, indeed, it is the
common doctrine of all thofe who coincide
with him in opinion, that the Houfe of Commons was always defigned to reprefent, not the
people at large, but the people of property, in the
kingdom. Admitting this to be a fact, let us
fee how far its prefent ftate agrees with the
defign. Are the paupers of Stockbridge, Barn-
ftable, Seaford, all the boroughs in Cornwall,
and a majority of the boroughs in the kingdom,
the men of property ? Why are the wealthy
Inhabitants of St. Mary-le-bonne, and other
parifhes in London, who in point of property,

I   are

are able to buy the electors of a majority of
the Houfe of Commons, excluded from fending Reprefentatives ? Why are the wealthy
Merchants and Manufacturers of Manchefter,
Sheffield, Birmingham, and even of London,
denied their right of voting ? But, it will probably be faid, a rich and commercial man, may,
at the expence of five thoufands pounds, procure
a feat in Parliament, if he defires it, and this
fhews the utility of the rotten boroughs, and
proves that property, predominates in the
Houfe of Commons : To this I afk, will a commercial, or any other man, lay out five thoufand
pounds without the profpect of a return ? If we
fee a rich avaricious man, expending immenfe
fums, to obtain the influence over a borough,
we cannot miftake his object ; it muft be with
him an adventure, a fpeculation, which he hopes,
will, in one way or another, return a propor-
tionate profit. And the mifchief is, that the
traffic may not be merely confined to Britifh
fubjects, but may, by foreign Princes, be converted into the means of ruling our Councils,
and ruining us as a nation ; for, if Mr. Pitt fpoke
truly, when he declared, " it was notorious
" that the Nabob of Arcot had fifteen Mem-
" bers in that Houfe, and that they did not act
" upon an identity of intereft with the people,"
why may not any other Prince fend in his members, and by expending a million fterling, take
the money out of Englifhmen's pockets, and
force them to fight his battles ? If it be faid,
though the Electors are poor, the Reprefentatives are wealthy, I anfwer, that we not only
find beggarly boroughs, but needy members
fitting

fitting for them. And as to " the property of the kingdom being in the hands of the mem-" bers *," I will afk where is the property (divefted of their places and penfions, and other political emoluments) of Meff. Pitt, Dundas, Jenkinfon, Long, Rofe, Steele, Addington, and Burke ? Thefe are the leading men in Parliament, yet, according to Mr. Young's argument, they muft lead to mifchief, becaufe they have little or no property.—Look at the oppofition :—See, (according to Mr. Young's argument again) what ample fecurity we have that they will act for the public good, becaufe they have large poffeffions. Meff. Grey, Whitbread, Lambton, Byng, Wharton, M. A. Taylor, Baker, Lord Wycombe, &c. &c. either poffefs, or are immediate heirs, to large eftates ; and Meff. Erfkine and Sheridan are of all others, interefted in the peace and profperity of the country, as their lucrative incomes from the bar and the theatre, would be the firft fpecies of property likely to fuffer from a convulfion.—Go up to the Houfe of Peers, and fearch for the property (places and Penfions excepted) of Lords Grenville, Hawkebury, Chatham, Auckland, and Loughborough. Compare the leaders of adminiftration and oppofition throughout, and it will be found, that the former are poor, and the latter are wealthy ;—Nay, take the aggregate of the number, and fortunes of " the Jacobin Society of the " Friends of the People," and compare them with the Houfe of Commons, (places and penfions excepted) and they will, perhaps, be found to be as wealthy and refpectable, as that pure body. If, as Mr. Young afferts, property were to be the foundation of our confidence in public men, there would inftantly be a change in adminiftration, and a Reform of Parliament. I therefore deny, that in the conftruction of the Houfe of Commons, there is any operating principle or controul, which fecures the protection of property, more than there is among a club of ftock-brokers, whofe primary object always is their own immediate advantage, whether it be connected with the welfare of the nation, or not. Hence, if we are to be guided, by what Mr. Young affirms, of the Houfe of Commons, that *it is a Reprefentation of property*, a Reform becomes as indifpenfible as upon any other principle. What, though fome members are men of property, they may confent to invade the poffeffions of the rich in general, becaufe they may receive a private compenfation that will not only reimburfe their own perfonal lofs, but plentifully reward them for betraying the interefts of men of property at large, who have no more controul over them, than the Swinifh multitude of Manchefter and Birmingham. Diftribute the elective franchife, equally among men of property, and I am content, becaufe I believe fuch a diftribution would render the Houfe of Commons independent of the Government, which is the true object of any Reform.

It may, perhaps, be faid, our Conftitution, originally, recognized only as men of property the owners of landed eftates ; that they all now

have

have votes, and therefore the ancient fpirit is
preferved. Without enquiring into the abufes
of County Elections, or fhewing the futility of
the expectation, that eighty county members,
could fecure us againft four hundred and
feventy-eight from cities and rotten boroughs,
I fhall content myfelf at prefent with afferting,
that the ancient fpirit, principles, and " *Practice*,"
of the Conftitution, recognized as men, who
had a right to be confulted in the National
Councils, all who were of CONSIDERATION,
whether by Landed Property, as at firft was
the cafe, or by Manufactures or Commerce,
which afterwards occafioned Citizens and Burgeffes to be called to Parliament, and which now are
of much more importance to the kingdom than
all its land. I fhall fhew this more fully when
I confider how far a Reform can be conftitutionally made. At prefent, I fhall only affert,
that according to the original principles, and
practice, of the Conftitution, and according to
Mr. Young's definition of what the Houfe of
Commons *is*, (a Reprefentation, not of Perfons, but of Property), a Reform is abfolutely
neceffary, becaufe it neither agrees with what
it originally was, and now ought to be, nor
with his defcription of it ; for Property is not
even fo much reprefented in the Houfe of Commons as Perfons.

Such, upon the principles of the Anti-Reformers, is the neceffity of reforming Parliament ; and if the practical evils arifing from
the defective Reprefentation are confidered,
they will be found ftill more urgent in favour
of that meafure. It is, I believe, univerfally
admitted,

admitted, that the prosperity of this country has arisen from the superior portion of Freedom enjoyed by its inhabitants, above those of the surrounding nations : If that be the case, the continuance of our prosperity must depend on the continuance of our Freedom ; and, if we find the House of Commons easily agreeing to narrow the latter, we cannot expect that the former will have a very long existence. The ready acquiescence to laying on additional burthens ; the profusion of the public money, and the invariable, enormous accumulation of debt, demonstrate the certainty, that at some time, we must reach the summit of the borrowing system : And even before this natural death of our public credit, the flames of some unnecessary War may melt the waxen wings, on which, like Icarus, we have towered to such an unnatural heighth. But what is more immediately alarming, the most arbitrary laws are found necessary, and are enacted, to ensure the collection of the Revenue : Those Laws are destructive of Freedom, and consequently must destroy our prosperity, for they openly and unequivocally invade our Liberties, and tend gradually to abolish and totally to annihilate them. The Right of an Englishman to a Trial by a Jury of his equals, has long been esteemed the dearest he possesses ; but this, in certain cases, which, with the increase of our debt, are annually accumulating, is compleatly abolished. It is not that justice is done between man and man that constitutes Liberty, so much as that justice is done between the Government and the People. In the most arbitrary and tyrannical

rannical Monarchies, ſtrict juſtice is generally adminiſtered between individuals; nay, it is more eaſy to be obtained, perhaps, than in Britain, becauſe the expence is leſs. All men act juſtly, unleſs they have an intereſt in acting otherwiſe: Deſpots, therefore, can have no intereſts in deciding partially between individuals: On the contrary, they will be eager to do rigid juſtice, in order to palliate their own robberies. Of what importance, then, is it, that we have Juries to try actions between John and William? Theſe might be as fairly decided, if there were no other Jury than the twelve, or four Judges. It is in caſes between the Crown and People that injuſtice is to be guarded againſt. Yet, we find Parliament yearly ſanctioning in the Exciſe, Stamp, and other Revenue Laws, the gradual Abolition of the Trial by Jury; becauſe, according to the late Earl of Chatham and Sir George Saville, their Corruption involved us in the American War, which brought a debt of an hundred Millions on our heads; and, according to Mr. Young, they are now to add another hundred to deſtroy a combination of Reformers; i. e. to preſerve to Lord Camelford and Egremont the privilege of ſending Members to Parliament for the Sheep Cots of old Sarum, and the Stones of Midhurſt! And to preſerve " Extravagant Courts, Selfiſh " Miniſters, and Corrupt Majorities!" (vide p. 171.)

The oppreſſive operation of thoſe Revenue Laws, particularly of the Exciſe, has often been ably demonſtrated. During the preſent Adminiſtration, they have been eagerly extended.

What

What can be more vexatious than that the Stamp Office fhould keep in its pay Informers, who, before they will be fuch, muft have loft all fenfe of fhame or honefty ; who muft have abandoned all hopes of ever being refpectable in fociety, and who, confequently, muft hate, and become the enemies of mankind, becaufe they know mankind defpife and deteft them ? What can be more vexatious or deftructive of Freedom, than that one of thefe reptiles may go into a tradefman's fhop, purchafe a pair of gloves, and by quirk or perjury, (for the oath of fuch a mifcreant, is fufficient, and is taken in preference to that of the moft refpectable tradefman) fine the vender in ten guineas, half of which goes to himfelf? Is this tried by a Jury? No. Who tries it? A Magiftrate, or two, appointed by, and receiving a falary, from the profecutor, (the Government ;) it is tried by men who fubfift on the fund to which the other half of the penalty is carried, and who, therefore, in a certain degree, have a common intereft with the Informer. For additional inftances of this kind, befides the Excife, Farming of Taxes, &c. confult the Hatters, Perfumers, and many other Shopkeepers. Liberty, which animates the induftry and enterprize of the people, and is univerfally allowed to have been the caufe of our profperity, can no more exift under a continual extenfion of thefe laws, than ice in a furnace ; and when once the caufe is removed, the effect will not long remain. Whether the Houfe of Commons, as at prefent conftituted, is likely to encreafe or diminifh the debt, which gives birth to thefe proceedings, is obvious.

obvious. Here I will not take theory, but Mr. Young's favourite " Practice," and according to practice, our debt will rapidly increase, and our Freedom as rapidly vanish. Spain, was once the moft powerful, becaufe the moft free nation in Europe : Her Cortes refembled our Parliament, but were, perhaps, much more pure. Corruption, indolence, pride, and op-preffion*, ftole into her government, from whence they were imbibed by her Nobles, and her People : Her Freedom, of courfe, gradually difappeared, until its bare remembrance was loft ; and behold now, how wretched are her inhabitants ; how infignificant as a nation !

Befides, the ready acquiefcence to Taxes, and to the arbitrary Laws found neceffary to exact them ; there are other circumftances, not lefs urgent in favour of Parliamentary Reform. Why were not the violators of the Conftitution in the cafe of the Middlefex Election punifhed ? Why was Lord Mansfield permitted to imprifon Bingley two years againft the Law ? Why were Minifters allowed to continue the American War without the fhadow of ability, or a hope of fuccefs, during feveral years, after the na-tion difapproved of it ? Or what is more recent, why was not an inquiry granted into the con-duct of Mr. Rofe, in the Weftminfter Elec-

* The moft deteftable means of her oppreffion, was, the Inquifition, which has been long proverbially execrated by every honeft Briton ; yet, Mr. Young fays, were he a Spa-nifh Minifter, he would not abolifh it ! Probably. he wifhes to introduce it into England, and to place it under the di-rection of Mr. Reeves, than whom not a more fit Director could be found.

K

tion ? Why have the chief agitators of the Manchefter, Mount-ftreet, and Birmingham Riots
efcaped juftice ? To thefe queries, Mr. Young,
will, perhaps, anfwer, " Why! Becaufe, all
" thofe meafures were only oiling the machinery
" of Government, to make it go fmoothly ;
" they were only fome of the oil of Influence,
" which is Corruption, in the eyes of Re-
" formers."

It is not neceffary to enumerate the penfion-
ing of Magiftrates, building of Barracks, and a
thoufand other fimilar inftances, in order to
convince every unprejudiced mind, towards
what centre the whole tranfactions of Parliament gravitate ; nor might it be fafe to animadvert on them freely. Let thofe, who defire a
more perfect illuftration, read the late Hiftory
of this Country, and particularly of the prefent
Minifters, where they may collect a fufficient
number to fill a volume. Shew me one inftance, during the boafted adminiftration of the
laft ten years, in which the Houfe of Commons
has fupported the interefts of the People, and
of juftice, in oppofition to the wifhes of the
Crown ? If it can be fhewn, that, upon one or
two marked occafions, it has done this, then, I
may think, its mifconduct rather the effect of
error, than of criminal acquiefcence : But if
this cannot be fhewn, then I muft either conclude, our prefent Governors to be the wifeft
that ever exifted in the world, fince they have
not erred once in ten years, or that Parliament
is fo fervile as never to thwart them. In the
cafe of the late Ruffian Armament, the Houfe
of Commons, having no fympathy of fentiment

ment with the nation, voted, according to the wifhes of the minifters, firft for, and then againft, a meafure. What was the confequence of this? They were proved to have acted contrary, and the Minifter afterwards to have acted in conformity, to the will of the people. By this means all the public efteem was transferred to Mr. Pitt, which would have fallen to the Reprefentatives of the people, if they had done their duty. Our Premier, in this way, has artfully gained all his popularity ; for, while he affects to defpife the will of the people, he, rather than hazard his fituation, obeys it on all dangerous occafions. And, as in the Ruffian Armament, he robbed the Houfe of Commons, fo in the prefent War, he robbed the Crown of the public affection ; for, when the calamities it brought on were daily rendering it more unpopular, his friends induftrioufly circulated a report, that it was then continued quite againft his inclination, but that higher powers (meaning the King and his friends) would have it fo. By this means, his popularity was at once preferved, both among thofe who wifhed to terminate, and thofe who wifhed to continue the War. It has hitherto been the practice, that when any Act of Grace was to be done, the King, by being made the inftrument in performing it, fhould reap all the advantage arifing from the national gratitude : Mr. Pitt, however, has reverfed this cuftom ; and when any odious meafure is to take place, the Crown is made not only the agent, but the parent of it, while his friends exculpate him, by fhruggs, and hints of regret, at the neceffity of fubmitting. Thus we

find

find him at once robbing both the King and Commons of the public efteem and gratitude, and throwing all that is unfortunate or obnoxious to their charge : He puts himfelf where the Monarch formerly ftood ; becomes the difpenfer of every thing that is gracious, and by this artful unconftitutional conduct, concentrates in his perfon all that is to be admired and beloved, either in the executive or legiflative power, and filches a popularity, not more unnatural, than formidable and alarming.

Notwithftanding the contumely with which the will of the people, calmly exprefied in an affembly of Delegates, is treated by the Anti-Reformers, yet they often juftify their own conduct on that very will which they affect to condemn : If it is faid, a Parliamentary Reform is neceffary, in order to afcertain the will of the people in a peaceable conftitutional manner, they anfwer, Parliament ought not to be guided by the will of the people, and therefore a Reform is needlefs : But throw the blame of the American War on Parliament, and they fhift their ground, they change fides, and boaft that it was popular. They juftify themfelves in commencing the American War, on the will of the people, and yet they deny that the will of the people fhould be their guide ! It is neceffary this fhould be decided ; Either let the will of the people, or the will of Parliament, fince they are to be diftinct things, be the rule of action. If the will of Parliament is to be the rule, then the blame of the American War attaches to them alone, and juftifies a Reform, in order to prevent another fuch evil : If the

will

will of the people is to be the rule, then, as the House of Commons voted directly againſt it, in the caſe of the Ruſſian Armament, that Armament juſtifies a Reform alſo.

But it will not be improper to enquire a little into the nature of the Will of the People. Upon examination, it will be found to be of two very diſtinct kinds, the one originating with themſelves, their own pure offspring, the other, a courtly baſtard. A clear inſtance of the latter may be found in the fate of Mr. Fox's Eaſt India Bill, which paſſed the Houſe of Commons without a murmur among the people, and was even approved, till the ſecret adviſers of the Crown raiſed an alarm about charters, and ſucceeded, in ſtimulating the public, to expreſs the ſtrongeſt diſapprobation of the meaſure. The popularity of the American War was alſo hatched in the Cabinet. The national indignation was inſidiouſly rouſed, by holding up the refuſal of the Americans to pay taxes, like ourſelves, while the true cauſe of the quarrel, the queſtion of Repreſentation, was kept in the back ground, or, if ſtarted, was anſwered, by the abuſes in our own ;—by inſtancing Mancheſter, Birmingham, &c.—by the deformity of the Britiſh Houſe of Commons, was the American War juſtified, and therefore, had a Reform taken place twenty years ago, one chief ſophiſtical pretext of that unhappy War, had not exiſted. It is likewiſe a courtly popularity, which has ſanctioned the preſent War. Proclamations, camps, and addreſſes, were not regarded with indifference ; Mr. Penſioner Reeves, from the Court, ſounded the trumpet of alarm, and

was

was echoed by the whole kingdom; thus a popularity, originating in the Cabinet, was given
to the War, which I call a courtly baftard popularity. But how was the American War
ftopped, or a War with Ruffia prevented?
Not by an expreffion of the public will, created
in, and directed by the Cabinet, but by an expreffion of the public will, emanating purely
from the people themfelves, and in direct oppofition to the wifhes of the Cabinet. On thefe
occafions, the people could not be duped; and
I think the will of the people, that fanctioned
in its commencement, and the will of the people that brought to a conclufion, the American
War, were as different in their nature as in their
object. Minifters will always be attempting to
create thefe baftard courtly wills, for their own
felfifh purpofes, to the injury of the nation; it
is the duty of the Houfe of Commons to detect
and expofe them, and to guard the people from
being duped into ruin; but, inftead of this, we
find the Houfe of Commons generally the chief
inftrument in promoting the delufion : in the
Ruffian Armament they were notorioouflyfo. And
therefore, to make them more careful of the
interefts of the people : to unmafk, rather than
cover the courtly defigns, a Reform has become
neceffary.

Thefe two different wills of the people were
to be traced in the fentiments and conduct of
the Alarmifts and Anti-Reformers laft winter.
They exulted in the univerfal loyalty of the
country, in the general affection of the people
for the Conftitution, and in their concurrence
in the meafures of government; they exulted

in

in the *courtly* will of the people. But when
the queftion of Reform was ftarted they deprecated the will of the people as the mother of
every calamity. If the War was confidered,
they vaunted that the will of the people was
in its favour;—if Reform was confidered, they
reprobated the will of the people as the moft
mifchievous of all guides. Was it ever boafted,
or given as an argument in fupport, of a meafure, that the will ot the people was againft
it ?—No.—Is not the reverfe almoft always the
cafe ?—Yes. Is not this proof of the neceffity
of knowing the will of the people, and confequently of Reform? Did not thofe who
boafted the general approbation of the War, and
the meafures of government, and thereby applaud the Will of the people, in fact confefs
the neceffity of Reform?—Nay, a more important confeffion is to be found in the defcription the Houfe of Commons gives of itfelf. In
impeachments, fuch as that now carrying on
againft Mr. Haftings, they profecute in the
name of ALL the Commons of Great Britain.
Thus they themfelves own, they ought to be
elected by univerfal fuffrage ; for to act in the
name of *all*, without confulting, or even of being capable of knowing the will of the *Majority*, appears to be inconfiftent with common
fenfe. And of this Mr. Young feems aware,
when he afferts, (p. 90.) that the Houfe of
Commons *are not* the Reprefentatives of the
People, and ought not to be fo called. Yet, it
was for this very fame affertion, that the Sheffield and Nottingham petitions for Reform were
rejected laft feffion of Parliament, and even
deemed

deemed libellous! Is it not, therefore, extraordinary, that Mr. Reeves fhould recommend a book, containing the fame affertions, which the Houfe of Commons deemed a libel, when contained in a petition? And is it not ftill more extraordinary, that the very men, who bellowed fo much about the dignity of the Houfe being infulted by thofe petitions, fhould reward Mr. Young with a place, for writing that, which, in St. Stephen's Chapel, they affected to condemn!

The addreffes, affociations, and general approbation of them, I confider to be direct confeffions, of the neceffity of Reform, becaufe they are confeffions of the utility of knowing the will of the people. The Houfe of Commons was inftituted for collecting and expreffing that will, and ought ftill to do it; a Houfe of Commons purely elected, being the only Conftitutinal mouth-piece of the people. For the government, therefore, to feek the fenfe of the country, in addreffes and affociations, and to pretend to be guided by the public will, when collected in that partial irregular manner, is, I think, highly unconftitutional, and a moft dangerous innovation, tending to fuperfede the ufe of the Houfe of Commons altogether. Nor, are our affiliated focieties, lefs eager to embark in all fuch meafures, than government are to encourage them. Acting in conformity to the fpirit which firft animated them, they ftill imitate the French; for, as the Convention decreed, on the 22d of November 1793, that no prieft fhould receive his penfion without producing a certificate of his having paid contributions of civifm; fo they have fet on foot civil contributions for flannel
waiftcoats

waistcoats, in all the public offices, to which those clerks, who did not subscribe, may see reason to believe, that they will be considered as *suspected persons*, and may even lose their places. The Archduke Charles has also published a proclamation, soliciting patriotic gifts, and voluntary contributions, from all well disposed persons, in the Netherlands, for the support of the War. Thus we find, not only the English affiliated societies, but the German Government, imitating the French mode of raising the supplies, and it may soon be as dangerous for a man in Britain, as, for a priest in France, to be without a certificate of his having made contributions of civism.

But, in considering how far our affiliated societies, associated for the purpose of protecting the Constitution, are themselves destroying it, we shall find, that addresses are not only contrary to the spirit of the Constitution, as they presume to speak the voice of the people, which can neither legally, nor fairly be done, unless in the House of Commons; but that those societies, in attempting to raise the supplies, have taken the actual, and most important part of the business of the House of Commons into their own hands, and if they succeed in their design, will render Parliaments totally useless.

The civic contributions for flannel waistcoats, were first begun in Edinburgh, afterwards in Windsor, and then the project was taken up in London, by a person notoriously in the pay * of Government, and in the confidence

L   of

* Besides having a share in the property of the newspaper, called the Sun, it is confidently said, he is allowed 600l. per annum

of perfons *, high in office. with whom' he daily communicates. A public meeting was next called, at the Crown and Anchor Tavern, where one of Mr. Reeves's committee, Mr. Devaynes, was appointed perpetual Prefident, and they refolved themfelves into a fociety for levying contributions of flannel waiftcoats, and money for purchafing them. The Common Council of the City of London, following their example, not only voted fupplies as a body, but nominated each of their members, tax gatherers,—collectors of voluntary contributions in their refpective fections; and Mr. Serjeant Watfon, in St. Andrew's, Holborn, opens his budget, with a poll-tax of five fhillings per head! It may, indeed, be faid, that thefe proceedings, though not ftrictly conftitutional, are of fo little importance, that they do not deferve the notice of Government; that the fums or clothes collected are fo trifling, compared with the neceffities of the ftate, that they never can fuperfede the bufinefs of Parliament in granting the fupplies: But we fhall find, that it is not the fault of thefe focieties, if their proceedings are not carried to the moft dangerous lengths: for, although they began with flannel waiftcoats, we find fuccefs has induced them to pafs to the providing of mitts, drawers, caps, fhirts, Welch-wigs, ftockings, fhoes, trowfers, boots, fheets, great coats, gowns, petticoats, blan-

annum for conducting that paper, and the True Briton, the latter of which is as confidenly faid, to be the property of perfons in office.

* Mr. Long, under Secretary to the Treafury, and Mr. Burgefs, under Secretary to the foreign department.

kets,

kets, &c. &c. From clothing the army, they have begun to victual them. A fchoolmafter, in the True Briton, recommends fplit-peas, as good for their health, and, therefore, fit to be provided by the affiliated focieties. From the army, they have proceeded to clothe the navy, and at laft, tired of going flowly on from one ftep to another, they exprefs their wifh of raifing the whole expences of the State, and thereby not only fuperfeding the ufe of Parliaments, but of financial Minifters ; not only of taking the bufinefs of the Houfe of Commons into their hands, but of taking the bufinefs of the Chancellor of the Exchequer alfo ! Mr. Chairman Bligh, and his affiliated fociety, in Chelfea *, declare their opinion to be, " that in " confequence of the public fpirit difplayed in " fubfcribing for extra clothing to the army, " *and confirmed of the value of our excellent* " *Conftitution,* they fuggeft that the whole " expences of the War might be defrayed in " the fame manner ; and, that in profecuting " the War, they earneftly wifh that the Mini- " fter, whofe office it is to find ways and " means for raifing the fupplies, could be re- " lieved from the political neceffity of impofing " taxes on the people." Thus they would relieve the Minifter, and the Houfe of Commons, from raifing taxes, to take the bufinefs into their own hands ! A propofition more hoftile to the Conftitution, and government of this coun-

---

* See the Advertifement, containing the Refolutions of this fociety, figned J. Bligh, dated from the King's Arms, Five Fields, Chelfea, and publifhed in the Sun, on Saturday, Nov. 30. 1793.

try,

try, than the imputed offences for which Meffrs. Muir and Palmer, are fentenced to be tranfported. Yet, the editor of the Sun, in a note to a correfpondent, about the fame time, remarks, not that fuch a plan is contrary to the laws of the land, but that he is afraid it would not be *fufficiently* productive !

The Hiftory of this Country will juftify me in afferting, that thefe proceedings are highly unconftitutional. Benevolences, or voluntary contributions, which at firft were free, were afterwards made compulfory, and were carried to fuch an extent, that the act paffed in the firft of Richard the IIId. " Damning and annulling " for ever that mode of raifing money," affirms, that thefe benevolences had been the ruin of many families, by obliging them to fell off their houfehold goods, and reducing them to beggary, &c. The attempt to raife forced benevolences, in a great degree, brought Charles the Ift. to the fcaffold, and fo fenfible was his fucceffor, Charles the IId. of this, and the danger of at all raifing money in that manner, that, in the fecond year after his reftoration, and in the zenith of his popularity, he did not prefume to do what a hired agent of Government, an editor of a newfpaper, now takes upon him to put in practice ; for we find, that in 1661, an act was paffed " to impower the King to receive " from his fubjects a *free* and *voluntary contri-* " *bution,* for his prefent occafions *,*" and the words of the act itfelf, not only exprefsly declare, that it fhall not be drawn into example

* See Rapin's Hiftory of England, vol. 2, p. 626.

for

for the future, but that *no aids of that nature,* can be iſſued or *levied,* but by conſent of Parliament *.

Now, the greateſt want of Charles the IId. was money to eſtabliſh a ſtanding army, and as the preſent " free and voluntary contributions" are alſo for the army, the object to which the ſums raiſed are applied, is, in a certain degree, the ſame: At any rate, King Charles the IId's occaſions for money could only be the general occaſions of the Government, ſuch as maintaining the army and navy, &c. and, therefore, the purpoſes for which the free and voluntary contributions were ſubſcribed to him, and the purpoſes for which thoſe are now ſubſcribed to Mr. Devaynes, are generally the ſame. The act, ſays, that no aids of *that nature,* ſhall be levied without the conſent of Parliament, and as I conceive, that a " *Free and voluntary pre-* " *ſent,*" whether paid into the hands of an Alderman, Common Councilman, a member of Mr. Reeves's committee, or an editor of a newſpaper,—or whether paid into the hands of agents, officially named by the King, as was done by Charles the IId. while it is paid generally for the uſe of Government, to be preciſely the ſame ; as I conceive, all of that kind, to be ſtill " Free and voluntary preſents," collected from the people for the *ſame purpoſe,*

---

* The Title of the Bill is, *An Act for a free and voluntary Preſent to his Majeſty.*

" And be it hereby declared, that no commiſſion or " aids of this nature can be iſſued out, or levied, but by con- " ſent of Parliament; and, that this act, and the ſupply " hereby granted, ſhall not be drawn into example for the " time to come." (See the Statutes at large.)

and

and as the act says, that no aids of the *nature* of a " free and voluntary prefent," fhall be raifed without the confent of Parliament, I think, I am perfectly juftified in affirming, the manner of now raifing them, is *illegal*. It cannot be denied, that they are precifely of the *fame nature* with thofe granted to Charles the IId. and there are the words of an act of the legiflature, ftating, that they cannot be levied but by the confent of Parliament. The prefent collectors of thefe aids can produce no authority for their conduct ; and if the Houfe of Commons were as tenacious of its privileges on this fubject, as it has fhewn itfelf on the fubject of petitions for Reform, the meafures it would purfue with refpect to thefe collectors of voluntary contributions, are obvious.

It is not only by the words of an act of Parliament, that we afcertain thefe voluntary contributions to be unconftitutional, but a principal member of adminiftration, the Prefident of his Majefty's Council, and a great lawyer, the Earl of Camden, gave it as his opinion, in the Houfe of Lords, during the American War, that they were highly fo: Mr. Dunning, and many other eminent characters, not only agreed with his Lordfhip, but ftrongly reprobated fuch a mode of raifing money for the public ufe. At that time, it was the fafhion to fubfcribe for raifing regiments and building fhips *, and when our prefent affiliated focieties

* The Earl of Londfdale engaged to build a feventy-four gun fhip, at his own expence, but his fincere attachment to the Conftitution, which he faw muft have been violated, had he fulfilled his promife, made him decline fo unconftitutional a meafure.

subscribe for the same purposes, which probably they soon will, they only then will have to pay the civil list, which their loyalty will induce them readily to do, and then bribery and corruption, at elections, may be prevented by never calling Parliament together, as there will be no occasion for its meeting.

I shall not dwell on the reproach which these contributions throw on government, whose more immediate duty it is to provide, not only necessary, but comfortable clothing to the army, than to give immense subsidies to Italian and German Princes, or to lavish rewards on the Alarmists ; and who cannot excuse themselves by saying, either the Parliament, or public, would not consent to pay the money : nor shall I dwell on the still greater impropriety, to give it no worse a name, of the members of administration, giving encouragement, through their agent, the editor, to so unconstitutional a measure. Its introduction is of recent date, and, I think, it ought instantly to be stopped, because, among other inconveniences arising from it, a very important one is, that three times the quantity of particular articles is provided, to that which is wanted, while, of other articles, not a third is subscribed. This is evident at present. Of flannel waistcoats, a sufficient number have already been contributed to give every soldier half a dozen, while the number of shoes would not afford them half a pair a-piece. Thus the public money is wasted, and the Constitution destroyed, by those who professedly associate to defend it. Parliament is the most proper monitor

of

of government; and a House of Commons, freely and frequently chosen, by a majority of the people, is the only constitutional channel, through which, to levy money, for the public use; because, in such a House of Commons, the voice of the people would be truly and calmly heard. On the contrary, addresses and associations, either for expressing their opinion on public affairs, or raising supplies for the State, are particularly dangerous, as the government pretends to be guided by the sentiments conveyed in them, though they admit of hearing only one side of the question, and may often be adopted to give an appearance of popularity to very unpopular measures, thereby deceiving the King, breeding disaffection in the people, and leading the one or other, or perhaps both, to ruin. Addresses and associations, therefore, should be discouraged, as delusive, and a Reformed House of Commons be substituted in their stead, where the will of the nation might be fairly and peaceably ascertained.

These are, I think, sufficient reasons to shew the necessity of reforming the Representation in this country. If we enquire how far it can be constitutionally done, we shall find, that the ancient spirit, and principles, and " *practice*," of the Constitution recognized *all men of consideration in the State*, as having a right to be present in the King's Councils. At the conquest, indeed, and for some time afterwards, men of landed property were the only persons of consideration in the kingdom; and accordingly we find, in the reign of Henry the IIId, in the first House of Commons of which we

have

have any account of thofe who compofed it, that it confifted *only* of four Knights from each county, who of courfe reprefented *landed* proprietors alone. The principles and " practice" of the Conftitution, however, being, that *all men of confideration*, who, confequently would be called on for pecuniary aid, fhould be prefent in the King's Councils, and trade beginning to fhew itfelf, Citizens and Burgeffes were fummoned from fuch towns as were growing into importance. And herein confifted the great excellence of the Englifh Conftitution: it adapted itfelf to all fituations, it opened its arms to all men of property as perfons more particularly interefted in the government of the country, and therefore entitled to a fhare in the Legiflative Affembly. But what an extraordinary deviation from the principle and practice do we now find? The men of the leaft confideration, a few of the pooreft beggars in the kingdom, create the majority of our Legiflators; while the men of moft confideration, of moft utility to the State, the merchants and manufacturers, are generally excluded. Conftruct the Houfe of Commons, according to the ancient principles and practice of the Conftitution, as manifefted by the admiffion of Citizens and Burgeffes to the National Councils; give every man, and none but men of property, a vote for Members of Parliament, and I am fatisfied.

Mr. Young (p. 223.) gives the authority of Dr. Brady, to fhew, that in ancient times, *liberi homines*, or free men, were only thofe who held in *capite*; and, throughout his whole book, he maintains, that landed proprietors were the only free-

M

reemen, and that they all had a right to vote
for Reprefentatives ; that the other inhabitants
were of no more account in the kingdom than
the cows, fheep, and hogs, they drove; the
artizans, manufacturers, labourers, &c. were all
flaves and villains, and the privilege of fending
Reprefentatives, was a gracious donation from
the King,—not a right of thofe to whom it was
given ; and in fupport of this he proves, that
feveral Monarchs called Members to Parliament
from obfcure places. I have already fhewn,
that it was their right, according to the fpirit
and practice of the Conftitution, which admitted all perfons of confideration to fend Reprefentatives ; and, it no more eftablifhes the
right of the Crown to beftow the privilege
where it pleafed, becaufe it did fo, than any
other improper act eftablifhes a right to do fo.
The felection of the towns as they grew into
importance, being entrufted to the Crown, was,
like many other duties, abufed by Kings, who,
apprehenfive of being in a minority in the Commons, fummoned members from fuch places as
they, or their creatures could command ; a defire to fecure, or bring into office, fome abject
favourite minifter, probably gave moft of our
rotten boroughs a right to fend Reprefentatives.
Some of our former Sovereigns, unconftitution-
ally, made Members of Parliament from fimilar motives, with thofe which induced his prefent Majefty, conftitutionally, to make fo many
Peers, in 1784 :—to fecure Mr. Pitt in power.

The affertion, refpecting tenants in *capite*,
when properly examined, is more deftructive to
Mr. Young's object, than any other pofition in
his

his book : For, admitting what he fays to be true, which, I believe it pretty nearly is, that none were freemen, but proprietors of land holding in *capite*, i. e. freeholders ; and that all freeholders, or freemen, had a right to vote in the election of Members of Parliament, it follows, that it is either a mockery to call the Englifh a " *free* people," " a *free nation*," as he does (vide the note, p. 205) or an injuftice, to withhold from them that right. He affirms, that formerly our mechanics, labourers, and manufactures, &c. were all flaves and villains,—that they were of no more importance than cows and hogs ;—admitted. But if they were flaves, were they free ? The queftion muft excite a fmile.—They were flaves ;—and if we are flaves alfo, then we have no right, upon the ancient principles of the Conftitution, as laid down by Mr. Young, to petition for Reform. —But where is the man, who dare come forward, and openly tell us, we are flaves ? What would be his fate ?—What would be the indignation of every honeft Englifhman ? Reeves and Young have indirectly told us this, but they have not had the audacity to fpeak roundly out. The fact is, that the prefent freedom of England was gradually extorted, fword in hand, from feudal fovereigns, deriving their rights from the fword of a conqueror ; nobly extorted. But had not the flaves and villains the greateft fhare in extorting it ? And fhall they who cemented it with their blood, be deprived of all his benefits ? While we are attempting to make freemen of the blacks in the Weft-Indies, fhall we meanly fuffer

fer to be proved flaves ourfelves? If it is true, as Mr. Young afferts, that in former times, none were freemen but freeholders, and that all freeholders, or tenants in capite, were invefted with the elective franchife, then it follows, that all freemen had a right to vote: If fix hundred years ago the majority were flaves, are we not now all called freemen? Is not the Negro, who, fixty years ago, was a flave in this country, called a freeman? And if every freeman had formerly a right to vote for Reprefentatives in Parliament, they who are now deprived of it, are, in that refpect, ftill flaves and villains. But our anceftors ftruggled for, and bequeathed us freedom, though not perfect freedom: The elective franchife was the principal article that efcaped them, and they overlooked it, becaufe, it was not then of fo much importance, nor was it fo much abufed as at prefent; they left to us the honour of atchieving it. Have we not equal virtue and perfeverance? Shall we not imitate their example? I am much furprifed that Mr. Young, in defcribing the Conftitution, did not difcover and recommend a mode of conftructing the Houfe of Commons, which would be highly fatisfactory to Government, which ftrongly refembles *his* theory of the ancient Parliaments, and is not quite unknown in the modern. As he afferts, that formerly none but tenants in capite, who held lands immediately of the Crown, had a right to fit in Parliament, or vote at elections, it might have occurred to him, that the ancient principles of the Conftitution would be revived, if, inftead of the right of election, and fitting in the Houfe of Com-

mons,

mons, being confined to thofe who held *lands*
of the Crown, it were given to thofe who held
places. Upon this plan of Reform, moft of the
Crown and Anchor committee, and even Mr.
Young himfelf, would be entitled to a feat in
Parliament.

" Hiftorians are agreed as to the Parliament
" of 1265, fummoned by an ufurper, being the
" origin of the Houfe of Commons," (p. 75.)
But (p. 216) Mr. Young, contradicts this; he
there quotes Sir Henry Spelman, and others, to
fhew, that in an hundred Parliaments *before* that
period, the *boroughs* never were reprefented;
implying thereby, that there *were* Knights of
the Shire, though no Citizens and Burgeffes,
which is exactly what I agree to, and fupports
what I have faid of the fpirit and practice of
the Conftitution, which called to the National
Council only men of confideration in the State.
It is true, that prior to 1265, and even then,
cities and boroughs were not reprefented, becaufe their inhabitants were of little or no importance; the landed proprietors were the only
men of importance in the kingdom, and therefore, the Houfe of Commons was compofed
wholly of their Reprefentatives. But as the
cities and boroughs became of *confideration*,
they alfo fent deputies. Nor is it juft to conclude, (vide p. 217) becaufe, during the 200
years after the Norman conqueft, the Houfe of
Commons was fo infignificant as not to be mentioned by hiftorians, that therefore, it never was
affembled: It would be equally juft to conclude, becaufe we have no account of the meeting of every county court, and other inftitution

tion of leſs importance, that therefore they did not at certain times aſſemble ; or, becauſe there is no hiſtory of the inhabitants in America, previous to its diſcovery by Columbus, that therefore, previous to that diſcovery, there were no inhabitants in that part of the world. And, even ſuppoſe the Houſe of Commons were never aſſembled, during the 200 years alluded to, that does not prove it to have been unknown to the Conſtitution, more than aboliſhing the Trial by Jury, in certain caſes, proves the Trial by Jury to be unknown to the Conſtitution. The exiſtence of Repreſentatives of ſome deſcription, may be traced in every page of our hiſtory, and is coeval with all law and government in England. What, although they were occaſionally laid aſide, their rights invaded, or their conſtruction varied, as ſuited the ambitious views of the Kings or Barons, that does not prove they had no right to aſſemble ? On the contrary, wherever we find the ſlighteſt traces of the Houſe of Commons, or wherever we find the moſt compleat deſcription and certainty of its exiſtence and power, there is no mention of its being a *new* inſtitution, which is ſtrong proof that it was an *old* one ; for is it poſſible, that ſo important a member of the Conſtitution could be created and introduced without ſome notice being taken of its novelty ?

Another inſtance, not only of how far a Reform can be conſtitutionally made in the Repreſentation, but alſo of its neceſſity, is the compleat change in its ſpirit and ſentiment. The Conſtitution, formerly, ſuppoſed a continual jealouſy of the Crown, and fellow feeling with the

the people, to exift in the Houfe of Commons;
but now, and efpecially by Mr. Pitt, it has been
made to repofe a continual confidence in the
Crown, and has fhewn, particularly on the Ruffian Armament, that it has no fellow feeling
with the people whatever. This dangerous
reverfal muft arife from the reverfal of the
mode of rewarding · the members, which is,
alfo, a great and pernicious change : for,
that Mr. Young's falutary Corruption, did
not formerly exift, is plain, fince the privilege
of returning members, now fo valuable, was
then rather a burthen than a benefit ; was then
of fo little importance, that many boroughs pe-
titioned to be eafed of it. To beget a jealoufy of
the Crown, and reftore the ancient nationality of
fentiment, in the Houfe of Commons, is, therefore, not only conftitutional, but the moft wife
and neceffary meafure, for preferving the Freedom and Profperity of the People.

How far it is conftitutional to fhorten the duration of Parliaments, does not require much
inveftigation, becaufe the facts relating to the
queftion are of recent date, and clearly underftood. Originally, Parliaments were only called
for a particular purpofe, and often fat only a few
days : Sometimes there were two, but generally
one new Parliament, in a year. In the feven-
teenth century, the ufage varied, and was moft
grofsly abufed by the long Parliament, at laft
diffolved by Cromwell. Yet, the vile precedent
was approved, and imitated by the Kings who
followed, and William the IIId. made it a great
favour to agree to the Triennial Bill. The Sep-
tennial Bill is juftified by Mr. Young, who fays
" The

" The Members of the Houſe of Commons, when elected, and in combination with the other branches of the legiſlature, aſſume and poſſeſs, and give themſelves ſuch powers and privileges, as rendered the ſeptennial act juſt as conſtitutional as the biennial." That act, however, was by thoſe who made it, juſtified only upon the exigency of the times, and ought to have been repealed when that exigency was paſt ; but, without conſidering its legality, I ſhall remark, that, if Mr. Young's pernicious doctrine were true, there would not be the leaſt ſecurity for the Liberties of the Nation : the three branches of the Legiſlature, in combination, might aſſume a power to repeal the Habeas Corpus Act, aboliſh the Trial by Jury, and the Liberty of the Preſs ; give to the King's Proclamations the force of Law, as was done in the reign of Henry the VIIIth. and veſting the whole executive and legiſlative authority in the Crown, diſſolve themſelves for ever, and annihilate at once the Conſtitution and Freedom of Britain : This, according to Mr. Young, they might *conſtitutionally* do, by *aſſuming* powers and privileges; and, indeed, he ſeems deſirous that they ſhould do it, when he ſets up as precedent, the example of Richard the IId. who dictated to the ſheriffs, the names of thoſe perſons whom they ſhould return to Parliament, and levied money without the conſent of Parliament : Richard's fate, as well as Charles the Firſt's, is well known. Were Mr. Young prime Miniſter, and his advice followed, he would probably occaſion events, which no good man can wiſh to think of.

On

On the fubject of the TIME for making a
Reform, I can fay nothing new. " To minds
" unwilling to do right, all times are equally
" inconvenient and improper. To him who
" diflikes the voyage, all the winds of Heaven
" are unpropitious : He looks for nothing but
" pretences to avoid it *." " This, indeed,
" is a never failing argument, equally in times
" of profperity and adverfity ; in times of War
" and Peace. If our fituation happens to be
" profperous, it is then afked, whether we can
" be more than happy, or more than free ? In
" the feafon of adverfity, on the other hand, all
" Reform or Innovation is deprecated, from
" the pretended rifk of increafing the evil and
" preffure of our fituation. From all this, it
" would appear, that the time for Reform
" never yet has come, and never can come+."
" When, indeed, the arbitrary monarchy of
" France, was battering down by the exertions
" of a great people, and nothing was feen but
" virtuous exertion and exultation, it might be
" admitted, that in fuch a conjuncture. men
" might run before the mark, and confound
" principles together, which had no connection.
" Such was the alledged, but not proved itate
" of England, when Mr. Grey gave notice
" laft year of his motion. The object i had
" then, therefore, at leaft, a *plaufible*, though
" not a juft foundation. But, good God ! how
" different, on the admiffion of the objectors
" to the times, was the prefent moment ?" the

* Vide Mr. Francis's fpeech, May 7th.
† Vide Mr. Grey's fpeech, May 6th.

Englifh

English " Starting back with horror at the
" crimes and calamities of France, and feem-
" ingly forgetting all diftreffes in an enthufiafm
" for their own Government! Surely com-
" mon fenfe pronounced that to be the hour
" for reformation, more efpecially when it was
" left to themfelves (the Houfe of Commons)
" to originate and to fafhion it. So far from
" being urged on by the people to go too far,
" they trod like men that feared the ground
" would break under them, and could hardly
" be brought up to the point which their un-
" derftandings dictated. Let them feize, there-
" fore, this happy and providential crifis, to do
" with popularity and fafety, what to fave their
" country muft be done at laft*." But, inftead of embracing this favourable opportunity,
every means is employed by Government to defeat the caufe of Reform, and calumniate its
advocates ; and the prefent mode of conftructing
the Houfe of Commons, will, probably, be continued until fome dreadful convulfion happens,
which may threaten the annihilation of the Conftitution itfelf.

In confidering what Reform fhould be made,
I fhall neither recommend nor reject any particular plan, being perfectly of opinion with Mr.
Grey, that, to conftitute the Houfe of Commons by univerfal fuffrage, or any other mode,
which would make it more independent than
at prefent, would be a moft falutary improvement. Mr. Young fays, the Society of the
Friends of the People, approve perfonal Repre-

* Mr. Erfkine's fpeech, May 6th.

fentation ;

fentation ; but I cannot difcover, except in his interpretations, that they ever have either approved or condemned it, or any other rule of Reform. Inftead, therefore, of enquiring what their plan may be, will it not be more in point to examine that propofed by the moft violent leader of the moft violent fociety in Britain ? —For fuch, I believe, Mr. Horne Tooke, and the conftitutional fociety, are deemed by Mr. Young, and thofe who think with him.

Mr. Horne Tooke, in his letter to Lord Afhburton, when he, and Mr. Pitt, Sir James Sanderfon, Mr. Froft, and the Duke of Richmond, were joint Reformers, fo far from approving univerfal fuffrage, recommends a plan of Reform, perfectly agreeing with the ideas of thofe gentlemen, who fay, the Houfe of Commons is a Reprefentation of Property for he makes it the governing power in electing the Reprefentatives of the people. He maintains, that although all men may have a right to a fhare, yet they have not all a right to an *equal* fhare in this choice ; for, fays he, " There is " a very great difference between an *equal* " *right* to a fhare, and a right to an *equal* " *fhare.* An eftate may devifed, by will, among " many perfons, in different proportions ; to " one five pou..ds, to another five hun- " dred, &c. each perfon will have an equal " right to his fhare, but not a right to an equal " fhare."

" This principle," (continues Mr. Tooke, alluding to univerfal fuffrage) " is further at- " tempted to be enforced by an affertion, that " *the all of one man is as dear to him, as the*

" *all*

" *all of another man is to that other.* But
" this maxim will not hold by any means ; for
" a ſmall is not, for very good reaſons, ſo dear
" as a great all. A ſmall all may be loſt, and
" very eaſily regained ; it may very often, and
" with great wiſdom, be riſked for the chance
" of a greater ; it may be ſo ſmall, as to
" be little or not at all worth defending or
" caring for. But a large all can never be re-
" covered ; it has been maſſing and accumu-
" lating, perhaps, from father to ſon, for many
" generations ; or it has been the product
" of a long life of induſtry and talents ;
" or the conſequence of ſome circumſtance
" which will never return. Juſtice and
" policy require, that benefit and burthen,
" that the ſhare of power, and the ſhare of
" contribution to that power, ſhould be as
" nearly proportioned as poſſible." Thus far
Mr. Tooke ſpeaks againſt the *equality of right*
to a ſhare in the Repreſentation : He then
ſpeaks of the impolicy of making the elective
franchiſe *univerſal.* " Freedom and ſecurity
" ought ſurely to be equal and univerſal," (ſays
he) " but the members of a ſociety may be *free*
" and *ſecure,* without having a ſhare in the
" Government. *The happineſs and freedom,*
" *and ſecurity of the whole, may even be ad-*
" *vanced by the* EXCLUSION *of* SOME, *not from*
" *freedom and ſecurity, but from a ſhare in*
" *the Government.*" Mr. Tooke then enu-
merates the claſſes which ought to be excluded,
and which certainly are the majority, as they
compriſe " the extremely miſerable," " ex-
" tremely

" tremely dependent," " extremely ignorant,"
and " extremely felfifh."

His plan of conftructing the Houfe of Commons is, that the kingdom fhould be divided
into 513 diftricts, each of which fhould fend a
member : that none fhould vote who were not
affeffed in two pounds to the parifh rates, or land
tax ; that Parliaments fhould be annual ; that
every elector at the time of giving his vote,
fhould pay two guineas, to be appropriated to
the ufe of the nation ; and, that where the
number of electors fell fhort of 4000, thofe
might vote over again, in proportion as they
were affeffed, and repeat their vote as often as
was neceffary to compleat the number of 4000,
ftill paying two guineas for each vote By this
plan, a landholder, paying a large land-tax,
might probably, have the privilege of giving
one hundred votes, upon paying two hundred
guineas. What plan would give the predo-
minating influence in the choice of Reprefentatives, more compleatly into the hands of men
of property than this ? According to it, property
would be *truly* reprefented, becaufe none but
men of *fome* property are affeffed to the amount
of two pounds to the parifh rates, or land-tax :
And even if univerfal fuffrage were eftablifhed,
yet the circumftance of paying annually two
guineas, for the privilege of voting, would effectually exclude the majority, who are poor,
and would ftill give the ruling influence in elections to men of property.

Such is the Reform propofed by Mr. Horne
Tooke, a leader, if not *the* leader, of a fociety
which Mr. Young defcribes to be much more
violent

violent than the Friends of the People ; and, if that fociety are called more dangerous than the latter, and yet it appears, the object of their leader is fo moderate, it is furely unneceffary to vindicate the Friends of the People from the charge of entertaining mifchievous views, fince Mr. Young admits them to be more moderate than the conftitutional fociety. I here think it neceffary to remark, that the epithet, violent, applied to the conftitutional fociety, is merely Mr. Young's, not mine.

Compare Mr. Tooke's plan of Reform with that propofed by a principal conductor of the prefent War, and confequently a partizan and favourite of Mr. Young, I mean the Duke of Richmond. His Grace contended for perfonal Reprefentation in its *fulleft extent*, and his co-adjutors * were Mr. Pitt, Lord Kenyon, and the leaders of the prefent Adminiftration ; yet it is ftrange, that the Duke and his friends, who firft recommended perfonal Reprefentation, and who truly have founded the focieties in favour of that meafure, and the petitions which lately prayed for it, efcape the cenfure of thofe who afcribe every mifchief to their doctrine ; and it is ftill more ftrange that fuch as Mr. Tooke, who are *decidedly againft* perfonal Reprefentation, fhould be loaded with obloquy !

From what I have ftated, therefore, it appears, that Mr. Tooke's plan agrees with the ancient fpirit and practice of the Conftitution,

---

* I do not mean to fay, that Mr. Pitt, Lord Kenyon, &c. contended for univerfal fuffrage, but as they acted cordially with thofe who did they are equally guilty, according to Mr. Young's reafoning.

which

which recognized only *men of confideration in the State* as having a right to a fhare in the great National Council ; and that the Duke of Richmond's plan agrees perfectly with Mr. Young's original Theory of the Conftitution, which gave every freeman a right to vote, and confequently, now, muft make the elective franchife univerfal, fince there are no flaves, but all are freemen in Britain.

For my own part, I think, the true object of a Reform is, not to give every man his natural right of a vote, but to make the Houfe of Commons independent of the executive power, or of a fmall number of wealthy men, and to make it act upon an identity of intereft with the people. The manner in which this would be beft accomplifhed, would, in my opinion, be the rule of a Reform. Mr. Fox faid truly, that the object of a Reform of Parliament ought to be the collection of the greateft number, not fimply of wills, but of *independent* wills : and Montefquieu was of the fame opinion ; for, in fpeaking of the Britifh Conftitution, he fays, it is neceffary in a free Government that every man fhould have a fuffrage who can be fuppofed to have *a will of his own*. I therefore agree with Mr. Tooke, that the happinefs of the whole may be advanced by the exclufion of fome, not from happinefs, but from the elective franchife, becaufe, a great manufacturer or landholder, or any other perfon, who employed, or could control, the extremely miferable, extremely dependent, or extremely ignorant, might influence them to vote as he pleafed, and thereby acquire an undue power in elections,

elections, and invade the independence of the Legiflature ; for I conceive, that, univerfal fuffrage, would render the miferable and felfifh electors liable to be corrupted, the ignorant liable to be mifled, and the dependent liable to be commanded. Thefe claffes would generally, if not always, form the majority at elections, and confequently, thofe candidates who poffeffed wealth, eloquence, or control, might procure themfelves to be returned, although their conduct in Parliament had previoufly been injurious to the national welfare, and even difpleafing to the independent national judgment.

It is afked, by Mr. Young, as the journeymen mechanicks, manufacturers, and labourers, never voted at elections, what right have they to petition for Reform ; and they themfelves may join in faying, if, according to Mr. Tooke's plan, we are not to vote, why, indeed, fhould we petition for Reform ? To this, I anfwer, it is as much their intereft to make exertions in favour of Reform, as it is the intereft of thofe who would be vefted with the elective franchife, which is, in itfelf, of no value to thofe who poffefs it, but like the delegation, ought to be entrufted to thofe who would exercife it with the moft wifdom and independence ; and as wifdom and independence prevail more certainly in the middling ranks, than in the whole mafs of the people, the elective franchife, fhould, in my opinion, be confined to them ; and I repeat, that it is as much the intereft of thofe who would not, as of thofe who would have votes, that a Reform fhould take place, and even that the elective franchife

-chife fhould be fo confined, becaufe the benefits that would flow from a Reprefentat.ve body fo conftructed, would be general, and the poor, as well as the rich, would equally partake of them.

If, indeed, men were to be guided wholly by natural right, and fully to infift upon the maxim, that no man is bound by laws to which he has not given confent, it would come to this, that all men in the kingdom muft affemble perfonally to form the Legiflative Affembly; and this was pretty nearly the cafe in ancient times. But the fyftem of delegation was adopted, not only for the fake of convenience, but becaufe, the wifeft and moft independent men, in whom the whole mafs could confide, were appointed to make laws, and by the fmallnefs of their number, were enabled to act with deliberation and found judgment. If, therefore, the ancient great National Council was thus narrowed, for the fake of convenience, wifdom, and independence, why fhould not the conftruction of the modern National Council be alfo, either narrowed or extended, to thofe limits which are moft likely to infure convenience, wifdom, independence, and impartiality? If the right is furrendered, in one inftance, to procure certain objects, why fhould it not be furrendered in another, if by that other furrender thofe objects are more firmly fecured ?

This will be further illuftrated by the Trial by Jury. The privilege of ferving on Juries, which certainly is of a more clear and immediate value, than a vote for a Reprefentative, is not. univerfal, but is confined to houfeholders

and

.t landholders, yet it is as much the natural
right of every man as the elective franchise.
And what is the reafon we never heard of petitions and affociations for extending to all men
the privilege of ferving on' Juries ? Becaufe all
men are fatisfied, that by the principle, it is fufficiently diftributed, (however, it may be fecretly abufed in practice,) to render Juries inde-
dendent and impartial ; and becaufe, the poor
labourer, or mechanic, although he knows he
can never ferve on a Jury, while he is neither
a houfekeeper nor a landholder, is, neverthelefs,
convinced that juftice is as amply fecured by thofe
who do ferve, as if the privilege were univerfal,
and therefore, he does not affociate or petition
to extend it ; yet he is as much interefted in
preferving it to houfeholders and landholders,
as if he himfelf were a houfeholder or landholder ; and if the elective franchife were as
generally diftributed, as the privilege of ferving
on Juries, we fhould neither have affociations
of the rich or the poor, for reforming Parliament.

But while I agree in opinion with Mr. Tooke,
that the middling clafs of fociety, fhould elect
the Members of Parliament, I am far from
agreeing with him in the mode of election.
There fhould be no fuch condition as that of
paying two guineas at the time of voting, nor
fhould any man be permitted to vote more than
once.

From what I have now faid, I think, a
Reform is neceffary ; that the elective franchife
fhould either be given *exclufively* to men of
confideration, or univerfally to *all men*, (for,

to

to me, it is immaterial, whether it is poffeffed by all, or a part, if the Reprefentative Body is independent ;) that the prefent time is peculiarly favourable for making a Reform ; that Parliaments ought to be annual, and that the Reprefentatives ought to be liberally paid by their Conftituents for their attendance. The means of preventing riots and corruption at elections, and making many other inferior regulations, are fimple and obvious.

Nor are we to dread innovation, for, while he deprecates the deftruction of " extravagant " Courts, felfifh Minifters, and corrupt Majo- " rities." Mr. Young fays, that " to declare " againft any meafure, becaufe, an innovation " is a conduct worthy of children. It is not " for or againft innovation, but what the na- " ture of the innovation fhall be." Admitting then, the removal of corruption by Reform, to be an innovation, is it not a more laudabie one than that which Mr. Young advifes, the eftablifhment of a national militia of property ? He cautions againft taking away from the machine of Government, a rotten wheel, which all men, not excepting himfelf, have directly or indirectly condemned, and yet defires the addition of a new one, which no man of eminence, has either approved or thought of. But he may be affured, his advice will not be followed by Government. The Crown will never part with the control over the military. During the American War, it found the confequences of putting arms into the hands of the people in Ireland, and although, men of property proftrate themfelves before the Throne at

prefent,

prefent, yet it is obvious, they will not always continue in the fame humour, becaufe, their property marks them out as the prey of taxation, and when their ridiculous fears are over, they will regard Government with a jealous eye; and if armed and difciplined, might probably make both Reforms and Revolutions; might deftroy felfifh Minifters, and root out Corruption. Men of property are not only more intelligent, but more interefted in the government of this country, than men of no property, and therefore are more likely to interfere in its conduct, and thwart its favourite meafures. It is not from an opinion of the wifdom of our rulers, that the rich at prefent fupport them, but from a dread infidioufly excited, that their property is in danger from the defigns of thofe who oppofe the Government. When this unfounded dread is paft, they will be equally difpleafed with " extravagant Courts, felfifh Mi- " nifters, and corrupt Majorities," for invading their property as they now are, with the Friends to Liberty, from a miftaken notion that they meditate its deftruction. The fame motive, which now makes them afraid to innovate, (the prefervation of property) will, hereafter, make them clamorous for political œconomy,—for Reform. Government knows this well, and therefore, it will depend for fupport on the " Slaves and villains," on the " Beggars with- " out a fhilling," rather than on the opulent, who foon might be difpofed to make elections for Members of Parliament, and dictate to Government, " at the point of the bayonet." If then, the Houfe of Commons, is at all a good
inftitution,

inftitution, and an evil does exift in it, are we not to afcribe the birth of that evil to the deviation from its principles, profeffions, and original practice? And how fhall we expel the evil but by bringing it nearer to thefe?—This is Reform; and the moment we deviate from it, a dread of innovation becomes juftifiable, left it may be carried too far. But no fuch dread can reafonably be entertained while mprovement is confined to the original practice, the principles, and prefent profeffions of the Houfe of Commons, becaufe a limit is marked out, beyond which it is impoffible for Parliamentary *Reform* to go.

Mr. Young fays (p. 81), that the people never had the power of changing the Conftitution without being in its perpetual exercife. This is like his other abfurdities; for it is plain the people *always* have the *Power*, though not always the *right*; and the " Practice" fhews the fallacy of what he would inculcate, that if they began Reform they would always be altering. " People are not fo eafily got out of their
" old forms as fome are apt to fuggeft. They
" are hardly to be prevailed with to amend
" the acknowledged faults in the frame of Go-
" vernment they have been accuftomed to.
" And if there be any original defects, or ad-
" ventitious ones introduced by time or cor-
" ruption, it is not an eafy thing to get them
" changed, even when all the world fees there
" is an opportunity for it*." The hiftory of all countries, particularly of this, fhews, that

* Locke's Effay on Government.

the

the people are never eager to change their form of Government. A memorable and recent inftance of this is on record. At the conclufion of the American War, the whole blame of which was thrown on the Corruption of the Houfe of Commons, and when as many approved of Reform as at this time dread it, fo fearful were the people of forcing Government, or fo little inclined to interfere, that there were not fo many fignatures affixed to the petitions, praying Parliamentary Reform, as there were laft fpring. If the people are naturally fickle, why did they not come forward on that occafion? The fact is quite the contrary to what Mr. Young affirms, and juft as Mr. Locke ftates it. Hiftory fhews us, that Governors have generally, if not always, altered Governments for the worfe, and that the people have always altered them for the better. I except the cafe of France, the experiment having neither had time nor opportunity to be made there; and indeed Mr. Young acknowledges, that the important tranfactions in that unhappy country have been effected, not by the people, but by the *terror* produced by a very *fmall minority*.

I have declined pointing out any of the abufes in the prefent mode of conducting elections, or fhewing the abfurd manner in which the elective franchife is diftributed, becaufe thefe fubjects, and feveral others, are fo ably treated in the Petition and Reports of the Friends of the People; nor do I think it neceffary to reply to all Mr. Young's ridiculous animadverfions on thofe publications. He fays (p 84.),

" You

" You ſtate the Parliamentary Influence of the
" Earl of Lonſdale, Lords Eliott, Edgecumbe,
" &c. you ſtate a fact; but (p. 81.) with
" this ſyſtem of influence, which ſeems Cor-
" ruption to the eye of ignorance, the liberties
" the people have been conſtantly improving;
" —we are happy under the Government of
" Influence, how then can it be bad?"—He
might juſt as well ſay, " You ſtate the crimes
of Barrington, Hubbard, and other highway-
men; you ſtate a fact;—but with their ſyſtem
of livelihood, which ſeems robbery in the eye
of ignorance, the liberties of the people have
been conſtantly improving. We are happy,
though contributions are made on Hounſlow-
Heath, how then can they be bad?" Corruption, as I have already ſaid, is a bad part
of a good ſyſtem, and ought to be removed. And,
though I by no means blame, or allude to the
Noblemen above-mentioned, yet, I have no
doubt, that tranſactions take place relative to
Parliamentary Influence, which I, being an ignorant perſon, deem Corruption, that not only
deſerve, but if publicly proved, would, for the
ſake of decency, be as rigorouſly puniſhed as
any offence committed on Hounſlow-Heath *.

. * I will not, however, be too poſitive in this remark, for I
was preſent when a very rich man of the name of Smith
was, on the teſtimony of ſeveral others, committed for
groſs perjury, by a Committee which was trying a conteſted
election for Exeter in 1791, before which he had been examined as a witneſs. I ſoon afterwards ſaw, by the Newgate
Calender, that he was impriſoned by virtue of the Speaker's
warrant, to take his trial at the Old-Bailey for perjury. But
by ſome means, which I could never learn, perhaps by Mr.
Young's ſalutary influence, he was ſpeedily brought before
the Houſe, reprimanded, and diſcharged; nor, although it
was then hinted, that he was yet to be puniſhed, has any pu-
niſhment been inflicted ! ! !

And we might, with equal juftice afcribe, our happinefs to the robberies committed there, as to the influence, which, according to Mr. Young, the ignorant think Corruption in the Houfe of Commons.

But this boafted happinefs is, I fear, much over-coloured. Thofe who can buy a three-and-fixpenny pamphlet indeed, the fhopkeepers and merchants of the City of London, the members of country corporations, and all who poffefs the means of living eafily, may well fay, they are happy; but they fhould not take upon them to anfwer for the whole nation *. Do thefe men comprife the whole people ? No. Not a tenth, perhaps not a fiftieth part of the people. Yet, as they are more intelligent, con-fpicuous, bufy, and noify, in the world, they certainly make a great appearance. My un-fafhionable idea of *the people*, however, com-prifes the Swinifh Multitude, as well as the men of fome property.

I fhall not enter into an abftract definition of happinefs. If it is prefumed, that acquiefcence in a ftate is a proof of its happinefs, then the Turks and Tartars, and the Weft-India flaves, are happy, and it would be impolitic to im-prove their condition or reform the Conftitu-

---

* A Farmer-General in Languedoc, who received as much money from the old French as Mr. Reeves now does from the Englifh Government, was told many years ago, when Reform might have been made without being followed by any mifchievous confequences, how neceffary it was to retrench and amend the mode of carrying on the Govern-ment; to which, like our penfioners, he anfwered, *Mais pourquoi changer ? Nous fommes fi bien.*—But why change? We are fo well.

tions under which they live. And it will be found, that what is deemed happiness in Britain is mere submission, arising, not from an enjoyment of the comforts of life, but from the ignorance of the lower class of people; an ignorance, which Mr. Young recommends to be perpetuated and increased; and, indeed, it is a sure way of bestowing what some men mean by happiness; for how can a man regret the want of that to which he has never been accustomed? Had the inhabitants of this country been always confined to such happiness, they might now have been no better off than they were under William the Conqueror, or than the present wretched natives of Morocco. The diffusion of knowledge has been a chief cause of the superior degree of happiness enjoyed by the British subjects; but this happiness cannot be interpreted to be acquiescence, or a blind and ignorant content, because then the wild savages of America are happier than the most wealthy classes in Britain. It must be interpreted, prosperity, or a superior degree of enjoyment in the necessaries and comforts of life, and in the civilized intercourse of society. By this rule, therefore, we must judge of it; for otherwise, we shall find the most wretched beings content in the midst of want, and the most wealthy and prosperous discontented in the midst of plenty.

If, then, we decide on the happiness of the people at large in Britain, not by what may chance to make them acquiescent, but by what rationally ought to make them content, we shall find their happiness to consist chiefly in the

P   assertions

affertions of thofe who really poffefs that, and in the ignorance of thofe who do not. That which *ought* to give content, and confequently conftitutes happinefs, is, I conceive, a plenty of good food and clothing, of all the neceffaries of life, to fuch a degree, as would make every man and his family comfortable. How far that is the cafe I will leave any man to judge, who can impartially, and with fome intelligence on the fubject, compare all claffes in his neighbourhood. He muft not confine himfelf to the Royal-Exchange, to Grofvenor-Square, or Mr. Young's parlour; but let him examine the large towns, the manufacturers, mechanics, and country labourers, and he will find an immenfe majority of the people, who are not fo well provided with the neceffaries of life, as the paupers in a work-houfe. Even in London, the moft wealthy fpot in the empire, I will venture to affirm, that a great majority are not comfortably provided with the common neceffaries of life; and that if thofe, who are comfortably provided, are compared with thofe who are actually ftarving, I believe, the latter will outnumber the former. So much for the general happinefs enjoyed by the people in this country. Mr. Young, when feafting on venifon and claret, with the Committee of Penfioners, in the Crown and Anchor Tavern, might well exclaim, " Are not we a happy people ?"—But were he to go into the regions of Spitalfields, where he might feat himfelf among many thoufands who want a morfel of bread to put in their mouths, and fay the fame thing, he would provoke the indignation of every fenfible man, and

even

even Mr. Reeves, I hope, would blufh at his impudence *.

But,

* It is rather fingular, that formerly the moft porapous eulogiums were always made on our government, and hap-pinefs, by thofe who were violating the principles not only of freedom, but of common juftice. The following extract from Henry's History of Britain, p. 183 and 4, vol. vi. quarto edition, is one ftrong inftance of this : " The King," (Henry VIII. in 1543) " had borrowed great fums from a " prodigious multitude of his fubjects, of all ranks, for the " repayment of which, he had given bonds and other legal " fecurities. The Parliament, very generoufly, made the " King a prefent of all the money he had borrowed from his " fubjects, and declared his bonds and fecurities to be of no " value. The King thanked his two houfes in the politeft " terms, for their generofity, and gracioufly accepted their " valuable prefent, while his creditors were left to confole " with one another, and, put up with their lofies, as well " as they could. *The preamble to this iniquitous ftatute, is one* " *of the moft extravagant pieces of flattery that ever was com-* " *pofed.* In it they give a mournful defcription of the con-" *fufion, poverty, diftrefs,* and *mifry,* of ALL OTHER NA-" TIONS, and drew a very *flattering picture* of the *riches,* " *peace,* and *profperity,* of England, *during his Grace's* " *reign.*"——Judge Jeffries, in the memorable trial of Lady Lifle, fpoke thus to the jury. "Befides, gentlemen, we cannot be " fufficiently thankful to our God, for the mercies we enjoyed " under that bleffed King (Charles the IId. for, we are to " confider, that we lived in all the affluence of peace and " plenty ; *our Lives, Liberties, and Properties, inviolably were* " *fecured ; every man fat fafe under the fhadow of his own* " *vine, and ate the fruit of his own labour.* And while our " neighbours fuffered the calamities of War, we were fur-" rounded with all the bleffings of Peace, and flept fecurely " under the government of a gracious and merciful King." ——The Lord Juftice Clerk, on the late trial of Mr. Muir, fpoke in fubftance precifely the fame. The only material dif-ference arofe from his natural *averfion* to inebriety, which induced him to convert Judge Jeffries' intoxicating vine into a fober fig-tree. He faid, " It requires no proof to " fhew, that the Britifh Conftitution is the beft that ever " was fince the creation of the world, and it is not pof-

fible

But, admitting the eulogiums on our happiness to be juft, will not their truth be the greateft aggravation of the conduct of thofe who pronounce them, and yet plunge us from this elevated felicity into all the miferies of the moft wanton calamitous War that ever afflicted Europe? In the fame proportion as they extol our happinefs, do they increafe their own criminality, by depriving us of it? And it is in vain to fay, that the War was neceffary, or, that it was provoked by the French, for though a majority of people infift on thefe two points, yet a candid examination of facts, will fhew, that they have no other foundation than the prejudices of thofe who believe them.

I have already fhewn, that hoftilities were refolved on by Auftria, long before they were meditated by France *, who did all in her power to

---

" fible to make it better: *for is not every man fecure? Does* " *not every man reap the fruits of his own induftry, and fit* " *fafely under his own fig-tree?"*
* The following extract is unqueftionable proof of the wicked defigns of the German Princes. It clearly eftablifhes what is maintained in the early part of this Pamphlet, viz. that the continental Defpots were the aggreffors in the prefent War. It gives reafon to fufpect, that the King of France, defigned by his flight, to put himfelf at the head of the invaders, as it fhews, that long before his attempt to efcape, he knew of the plot, yet concealed it; and it completely refutes the affertion in a late Pamphlet entitled, " Reflections on the propriety of an immediate Peace," faid to be written by Mr. Vanfittart, that the Convention at Pilnitz, was only formed in confequence of the imprifonment of Louis, and that when he was fet at liberty, it ceafed to exift, becaufe, it fhews, that the plot againft France, was formed when the King enjoyed more indulgence than at any other period during the Revolution, and when the new Conftitution of France, wore the moft aufpicious afpect. M. Bigot
de

to prevent them. And although Mr. Young
afferts, that " this country had no right to in-
" terfere in the affairs of France, previous to
" the 10th of Auguft, and that till then our
" Government was rather friendly than other-
" wife," yet it will eafily be fhewn, that he is
quite miftaken, unlefs he means fuch friend-
fhip as the Duke of Brunfwick's, and that the
War might, and ought to have been avoided,
as the French were even more defirous of keeping peace with Britain, than they had been to
prevent a rupture with Auftria.

Our Government, indeed, obferved a ftrict
neutrality previous to the 10th of Auguft, but

de Sainte Croix, who was Minifter of Foreign Affairs, to
Louis the XVIth. at the time that Monarch was dethroned,
and who is now an emigrant in London, has, fince his arri-
val here, publifhed a Hiftory of the Confpiracy of the 10th
of Auguft, in which he fays, p. 152, " Dês le *Printems* de
" 1791, le Roi empêchoit l'execution d'un plan fecret arrêté
" a Mantoue pour attaquer, deux mois après la France, dont
" les armées etoient alors incomplettes, et les frontiéres fans
" défenfe."

" In the fpring of 1791, the King prevented the execution
" of a fecret plan, determined on at Mantua, for attacking
" France, two months afterwards, the armies of which
" were then incomplete, and the frontiers defencelefs."

This gentleman's authority is of the higheft nature. He
now avows himfelf always to have been a determined Royalift. With a laudable love of his King, however, he
thought he could moft fincerely ferve him by difguifing his
fentiments, and remaining about his perfon, which he did as
long as it was fafe to do fo. He was a particular confident
of Louis the XVIth. and he now abufes the Conftitution of
1789. But what is ftronger proof of his being a Royalift,
and a confident of his late King, is, that he is admitted to
St. James's, and careffed by our Government ; and, even
the well known Peltier, in his Dernier Tableau de Paris,
calls him " le veritable homme du Roi."

not from any good will towards France. It
was the dread of the refentment of the Britifh
people, and the belief, that the Duke of Brunfwick would effect his purpofe without our
affiftance, that prevented Adminiftration from
openly joining the concert of Princes in their
firft operations. Sufficient proof of this was on
record before the 10th of Auguft, and fubfequent events have fet the fact beyond all difpute. The hoftile difpofition of the Court of
St. James's, towards the French Conftitution,
was believed all over Europe, and never even
queftioned in England. The French knew and
avowed it; but they relied on the love of Liberty inherent in Englifhmen, for defeating the
defigns of Government againft their Freedom :
And for a while, this notion of a difference of
fentiment, between the King's Minifters and
the people, feemed to be juftly founded.

Mr. Burke, denounced the French Revolution previous to the abolition of titles, and before fo much blood had been fhed as lately was
fpilt in Briftol, about the payment of a half-
penny. He has continued his furious Anathemas,
and long before the 10th of Auguft, was feafted
at the cabinet dinners of our Minifters, and careffed by that Monarch, whom he had declared
the Almighty to have hurled from his Throne,
whofe houfhold expences he had curtailed, and
whofe difpleafure was well known, in confequence, to have been incurred. What then
was the caufe of Mr. Burke becoming fo great
a favourite, excepting his abufe of the French
Revolution ? Was this a fymptom of our
Government

Government being rather friendly than otherwife?

In the fummer of 1791, when our Government knew that the German Princes were
planning a War againft France, did not a newfpaper, notorioufly in the pay of certain perfons,
high in office, teem with the moft artful falfehoods and grofs calumnies, for the purpofe of
deterring the admirers of the French Revolution from celebrating that event on the 14th of
July, or of producing a riot which might dif-
grace the meeting? Did not a mob affemble
round the Crown and Anchor Tavern, in London, which, thanks to their own difcretion and
good intentions, rather than to thofe who attempted to dupe them into violence, difperfed
without doing any mifchief? Did not a fimilar
mob affemble in Birmingham, which committed the moft horrid exceffes for the honour of
Church and King? Was not that mob inftigated
by perfons, who fuppofed they were pleafing and
ferving the Government, and do they not yet
remain unpunifhed? But from thefe events,
can any thing be difcovered in the conduct of
our cabinet that fhewed it to be " rather friend-
" ly to the French Revolution than other-
" wife ?" A politician, would have formed his
opinion of the fentiments of our Government
from the contents of fuch a newfpaper, becaufe,
it would not dare to take a fide on a queftion of
fuch magnitude and continuance, *againft* the
will of its patrons; and, through its channel,
Adminiftration might inculcate fuch doctrines
as it might be impolitic to avow. In fuch a
newfpaper, they fpeak in the dark; they affert
what

what they pleafe, without being refponfible, or known, or even perhaps, fufpected. The pub-lic look for the printer's name at the bottom, to fee who is the author of the contents, when, probably, it would be more juft to look to the vicinity of Whitehall.

Mr. Pitt himfelf, in' the debate on Mr. Grey's notice of his motion for a Parliamentary Reform, three months before the 10th of Au-guft, called the French Conftitution, fuch, as, if formed in the morning, could not exift till noon. He alfo reprobated the wild French theories, which, he faid, were fubverfive of all order and government; and although he did not preach War againft France, yet he coun-tenanced thofe who did, and condemned as dan-gerous every principle of the Revolution : Nor did any member of Adminiftration ever hint that they difapproved of Mr. Burke's war-whoop.

Was the conduct of our Minifters in declining to negociate between France and Auftria, in favour of Peace, when folicited to that effect by the former, no proof that they wifhed to fee a War? Was the balance of power in danger by the Emprefs of Ruffia's retaining poffeffion of Oczakow, and in no danger by the com-bined armies gaining poffeffion of Paris, for fuch was then the expectation? Could we go to War about a fingle remote town, and yet re-fufe to negociate, when the exiftence of a great neighbouring nation was in queftion? " But " we had no right to interfere, unlefs called " upon by all the parties *."—No! Then we

* Vide Lord Grenville's note to Monf. Chauevlin.

have

have no right to interfere in behalf of Poland,
unlefs called upon by the Emprefs of Ruffia ;
then, we had no right to interfere about
Oczakow, in behalf of the Turks, becaufe,
the fame gracious Dame did not call upon us ;
and we have no right now to interfere in behalf of the French Royalifts, becaufe, the Republicans do not call upon us! We had a right ;—
it was our duty. We might have preferved a
limited Monarchy in France, averted the dreadful calamities of the prefent War, and faved an
amiable unfortunate King from the fcaffold.

Another pointed inftance of the fecret hoftility of our Government, towards the new Conftitution of France, is to be found in the cor-
refpondence which paffed between Lord Grenville and Monf. Chauevlin, in May, 1792. The
latter, invariably ftiles Louis the XVIth. " King
" of the French," which was the title decreed by the National Affembly, and the former
as invariably ftiles him, " His moft Chrif-
" tian Majefty," which was the title during
the Defpotifm. Thus, the Britifh Government,
three months before the 10th of Auguft, fully
manifefted its unwillingnefs to acknowledge the
new Conftitution : It was manifefted in the
fame manner, as was afterwards openly declared, on refufing to acknowledge the Republic, in diftinctions of names and forms ; and
betrayed a fympathy, at leaft, if not an actual
connection, with the Duke of Brunfwick. If
to this, we add the uncontradicted paffage in
the Declaration of the French Princes, on the
10th of September, 1791, wherein the King of
France, is " affured that *every* power in Eu-

Q   " rope,

" rope," (among which, Britain muft be in-cluded). " is favourable to the enterprife of the " Duke of Brunfwick," and the very warm reception, both at Court and by Minifters, of that public plunderer, Calonne, it would be the height of folly to queftion the regret of our Cabinet, at the fall of the old French Govern-ment, and the ardent defire for its reftoration.

In addition to thefe proofs, which exifted pre-vious to the 10th of Auguft, of the unfriendly difpofition of the Britifh Court to the French Revolution, others have occurred fince that pe-riod, which illuftrate and confirm my opinion of the fecret hoftility of our Minifters, and fully contradict Mr. Young's affertion, that, " till then, our Government was rather friendly " than otherwife."

. Mr. Pitt, declared, in the Houfe of Com-mons laft winter, " That as a right hon. Gen-" tleman, (Mr. Fox) had rejoiced at the re-" treat of the Duke of Brunfwick, fo he, on " his part, would fay, he confidered it as the " greateft misfortune that could have befallen " mankind." Lord Auckland, in his Memo-rial to the States General, dated Jan. 25th. thus expreffes himfelf, when fpeaking of France, " It is not *quite four years* fince certain *mif-* " *creants*, affuming the name of Philofophers, " have prefumed to think themfelves capable " of eftablifhing a new fyftem of civil fociety." Lord Hood, in his firft proclamation to the people of Toulon, not only has imitated Lord Auckland, in alluding to " the mifery which " for *four* years has afflicted France," but de-" clared, " that it is for the *re-eftablifhment*
" *of*

" *of the French Monarchy*, that Britain has
" armed." But, indeed, the late Proclamation of his Majefty, fets the fact beyond difpute, that we are fighting to give a Government to France, and is therefore, ftrong pre-
fumptive proof of the hoftile difpofition of our
Court, previous to the 10th of Auguft, and,
combined with other circumftances, fairly juftifies us in concluding, that it was only the fear
of difcontent at home *, and the confidence that
our exertions were unneceffary, which prevented
us from being more early engaged in the War.

Much proof cannot be required of the unfriendly difpofition of our Court after the 10th
of Auguft. Withdrawing our Ambaffador, and
refraining to fend another, was an unequivocal
demonftration of difpleafure, and a fure préfage
of hoftility. The fcandalous and inceffant abufe
of Monf. Chauevlin, in a newfpaper, (the Sun)
conducted by perfons connected with, and at
the devotion of Adminiftration, was furely fome
ground for fuppofing there were thofe who
wifhed to drive him from this country, and
thereby precipitate a War. The French, on
the other hand, ftrongly evinced their defire to
keep Peace with this country, by continuing
Monf. Chauevlin at our Court, after the Britifh
Ambaffador was withdrawn from Paris : This
was an inftance of humiliation, which, even
amidft all their fucceffes, exultation, and pride,

* Mr. Burke, a few days before the declaration of War,
faid, in the Houfe of Commons, that he only pardoned the
Minifters for their flownefs in beginning the War, on account of the neceffity of waiting till the public temper was
inflamed to a fufficient pitch to fecond them effectually. For
nearly four years he had been waiting for that fortunate period, which had at laft arrived.

they fubmitted to, in hopes of maintaining tranquillity: And among many others, equally unqueftionable, may be enumerated, the refufal of giving permiffion to Dumourier, to enter Holland, who, in a council at Liege, on the 5th of December, affirmed, that he could eafily march to Amfterdam, and deftroy the Dutch Government, if he received orders for that purpofe. The Executive Council, however, would not give him any fuch orders, for 'fear of provoking a rupture with Britain; a conduct, which they, no doubt, foon repented, as they faw by the ftoppage of corn, the alien bill, and the fpeeches of Minifters in Parliament, how unfounded was their expectation of continuing Peace: They faw, that the antipathy of the Britifh Government to their Revolution, which had been manifefted from the beginning, and had gradually increafed and difplayed itfelf, was at laft, going to break out into open War. The King's death gave a plaufible pretext for difmiffing Chauevlin, and by provoking the indignation of Englifhmen, prepared their feelings to plunge into a War of vengeance; for fuch it certainly was on the part of the Britifh people, even in its outfet.

With regard to the profeffed grounds of commencing the War; the decree of fraternity, the opening the Scheldt, and the aggrandizement of France, they have all been fo amply difcuffed, that it would be fuperfluous in me to animadvert on them; and, as was often faid in Parliament, there can be little doubt that they might have been amicably fettled, if a pacific difpofition had been manifefted by our
Minifters,

Minifters, and if they had entered into a proper negociation. They indeed, pretended, that the French had given in their ultimatum, but our Minifters have fhewn, that when it fuited their purpofe, they could confider ultimatums only as preliminaries, for, in the late difpute with Spain, they appeared as anxious to avoid, as, with France, they fince appeared, eager to precipitate a rupture.

Such were the pretexts for commencing the War : Its * real grounds, the fubfequent conduct of Minifters, has fully illuftrated ; they have fhewn that Mr. Young was right in faying, its object was to deftroy a combination of Reformers. Thus, then, according to the late Earl of Chatham and Sir George Saville, the American War was begun in order to gratify the Corruption of the Houfe of Commons, and according to Mr. Young, the prefent War is to preferve its rotten Conftitution. Indeed, Mr. Young's opinion is confirmed by every circumftance ; for, was not the late alarm directed wholly againft Reform ? Was not the proclamation, in May 1792, produced in confequence of the fociety of the Friends of the People being inftituted for the purpofe of procuring a Reform ? What could be the object of that proclamation, if it was not to excite an alarm againft Reform ? Did it not immediately divide the kingdom into Reformers and Anti-Reformers ? Was it not the avowed determination of

---

* Mr. Bowles, one of the Crown and Anchor committee, in his " Real Grounds," publifhed laft winter, fays, it is merely on our part a War of *Defence*, and that no country has a right to interfere in the internal concerns of another.

Government to refift Reform, and the dread, that in the conflict, the Conftitution would be deftroyed, that firft founded the alarm ? It was from the fpeeches of thofe in power, from proclamations, camps, and addreffes, that the kingdom firft began to think itfelf in danger. In the fame manner as the popularity of the American War was planned in, and directed by the fecret Cabinet, was the late alarm planned and directed by the tools of Government. Alarms have always been found ufeful to thofe who profit by " extrava- " gant Courts, felfifh Minifters, and corrupt Ma- " jorities." The alarm excited by the riots in 1780, deftroyed all hopes of a Reform at that time ; and the alarm at the conclufion of 1792, has again defeated, for the prefent, the fame caufe. Thus we find, that alarms will always be hatched, when Parliamentary Reform is likely to fucceed, and yet, that the want of that Reform—brought on the American War, the moft ruinous this country ever faw, and has involved us in another, the confequences of which cannot be calculated. although they threaten to be much more dreadful.

After a period was put to negociation, excepting fome underhand intercourfe, which, it would appear, Minifters entered into merely to give a colour to an affected defire of Peace ; after the difmiffal of Chauevlin, what was to expected ?— War.—His difmiffal was an unequivocal mark of hoftility on our part, for fo fuch a ftep has always been confidered by nations in fimilar circumftances. The office, then, of commencing the War was thrown upon the French, who, it was not to be fuppofed, amidft their fuccefs, would

would betray a daftardly fear, by declining it:
It was not to be expected that they would then
fhrink from a rupture with a country, which
fhewed it would begin War as foon as it could
with advantage, and which had cut off all chance
of the continuance of Peace, by driving away
the Ambaffador. Yet Mr. Dundas had the af-
furance to declare, in the Houfe of Commons,
that the War, on our part, was fimply defenfive ! A declaration, which he could only have
been encouraged to make by the ready belief
then given to whatever was faid in fupport of
Government. He might as juftly have declared,
that the feizure of Poland was fimply defenfive
on the part of Ruffia and Pruffia ; and indeed,
thofe powers made that their pretext ; they faid,
they divided Poland in order to defend themfelves againft Jacobinifm !

War, however, being commenced, it is not
fo important at prefent to inveftigate its original
pretexts, as to enquire what are now its real objects, how long it is likely to continue, and
what will probably be its termination and confequences. Upon the firft of thefe points, there
are as various opinions as there are about religion : Some are for fighting to reftore the ancient Monarchy, and exterminate the prefent
popular principles ; others, defire the reftoration of the Conftitution, founded in 1789 ; a
third party, fupport the War in hopes of gaining territory ; a fourth, wifh the throat of every
Frenchman may be cut ; and a fifth, fupport it,
becaufe—they hate the French. Mr. Young,
though not fingular, differs fomewhat from all
of thefe ; he fupports the War, in hopes of deftroying

ftroying a combination of Reformers, and procuring Peace for the next fifty years.

To thofe who would continue the War, in
hopes of reftoring the ancient Monarchy, which
certainly cannot be done, without deftroying
the prefent popular principles, I fhall obferve,
that a War againft opinions never was fuccefsful. Perufe the hiftories of the Wars againft
religious Opinions, againft religious Reformers,
againft the Principles of Freedom in Holland,
againft the Principles of Freedom in America.
Were any of thefe fuccefsful, were any of the
fame kind ever fuccefsful? No. In the third
campaign, the caufe of the Americans, feemed to
be quite hopelefs, twenty times more fo than
that of France now is, yet it ultimately triumphed.
Are not the French the moft powerful Nation
in Europe? Have they not formerly, when
lefs interefted in the caufe, contended againft and
repelled all Europe? Are they not now animated even to madnefs with hatred againft
Kings and Nobility? Does not every fuccefs
of the Combined Armies encreafe the unanimity, and confequently the ftrength of France?
If enthufiafm in La Vendee has fo long refifted
that power, which fo lately threatened the conqueft, and fince, has fuccefsfully withftood all
Europe; if in one department it has coped
with the competitor of Germany, Italy, Britain,
Spain, Holland, &c. what may it not be expected to do in eighty departments? We are
led, indeed, to believe, that bribery will produce
revolts, and divide the French; but this is the
very fame expectation which was held out
during the American conteft, and then much

more

more fuccefsfully practifed than hitherto has been done in France. The fame men are purfuing the fame policy. But Hawkefbury, Dundas, Howe, Auckland, and Loughborough, may have no more reafon for exultation than they had in 1780. Bribery may, indeed, do much mifchief, and it is natural for thofe who believe it perfuafive, to try its effects on others. But " extravagant Courts, felfifh Minifters, " and corrupt Majorities," are not " inti- " mately interwoven" with French Freedom, and therefore cannot be expected to produce profperity and happinefs in that nation. They do not fight for the glory of a Court, or the folly or ambition of a Minifter ; every man thinks he fights for himfelf. Even the Tou- lonefe, while furrendering their town, declared. their firm attachment to the Conftitution found- ed in 1789, and the reftoration of that Con- ftitution is certainly the furtheft ftretch from Republicanifm to which the French people will confent to go. It therefore follows, that either the allies muft relinquifh their original views, or animate the French nation, as one man, to maintain the conteft againft them. And here I will appeal to Mr. Young's favourite " Practice" and " Events," and defire to know, if a War againft a people, againft fo powerful a people as the French, has ever been fuccefsful ?

To thofe who wifh the reftoration of a limited monarchy, of the Conftitution founded in 1789, it cannot be neceffary to fay much to convince them of the hopeleffnefs of their object. They are, indeed, the only reafonable clafs, and it is

there-

therefore lamentable, that they should have the least prospect of success. It was that Constitution which Mr. Burke reviled, which Mr. Pitt condemned, and which, as I have already shewn, never was sincerely approved by our Government. It is that Constitution which Lords Auckland and Hood attempted to hold up to detestation, when they mentioned the miscreants, who, for *four* years, brought misery on France. But what is more, it was against that Constitution, before it was even finished, that the German Despots made War, and to acknowledge it, would be to acknowledge they had failed in their design, that they were defeated. If further proof is wanted, that the Constitution of 1789, and the present Government of France, are equally odious to the continental Despots, look into the prison of La Fayette, who attempted to fix that Constitution, and to save and support the King:— Robespierre or Hebert, if, in their power, could not be treated with more cruelty.—Nay, it is even vain to expect, that the allies design the restoration of the *genuine*, the *ancient French Government*! The States General was a part of the ancient Government, and it was the States General that brought about the Revolution, and framed the Constitution founded in 1789, and accepted by the King in 1791. The Convention at Pilnitz was formed against the States General, after it had given itself another name, indeed; and Lords Auckland and Hood have called them miscreants, who have brought misery on France. It was the States General which began the Revolution,

and

and, according to the allies, began all the
prefent mifchief. It cannot, therefore, be the
reftoration of the *ancient* Conftitution which is
intended, it can only be a crippled Defpotifm.

To thofe who confider the objects of the
War, to be the acquifition of territory, it may
be obferved, that all which we fhall probably
conquer and retain, will, by no means, com-
penfate for the expences incurred by the continuance of hoftilities ; and even acquifition of
territory, may be of no folid advantage, if
we are to believe the fpeeches and writings of
the fupporters of Government, 'who have inferred, from the national profperity of the laft
ten years, that the lofs of America has been
rather a benefit than an injury to Britain. To
perfons who confider conqueft as the object of
the prefent ftruggle, I will not remark, on the
infamy of making War for plunder, becaufe,
with them, and the tyrants of the continent,
fuch a remark would be treated as a jeft.

Thofe perfons who would continue the War,
merely becaufe they hate the French, and hope
to cut their throats, I confign over to the Crown
and Anchor committee, hoping there are
fufficient humanity and religion, in that worthy
body, to make them blufh at their brutality,
and tremble at their breach of the laws of God.

Mr. Young, is for continuing the War, to
deftroy a combination of Reformers, and give
us fifty years Peace ! This fhews the depth of
his penetration. I appeal to all the " Experi-
" ments," " Practice," and " Events," that
he can produce, whether all Wars have not,
inftead of deftroying, created Reformers. As

the

the people feel the weight of burthens, they
begin to think of lightening them, and confequently, the firſt thing they turn their thoughts
to, is Reform. Was not this particularly the
cafe of the American War? What produced
ſuch a combination of Reformers in 1780, if it
was not the expences of that conteſt, and the
miſmanagement of the public purſe? The fame
effects, will reſult in time, from the prefent
War. Every new tax will make Reformers of·
the claſs which is fixed on to. pay it : And even
Mr. Young himſelf, may again think a Reform
defirable, if the neceſſities of the State ſhould
oblige Government to diſcontinue his ſalary,
and encreaſe the land-tax.

If he could give any proof in fupport of his
aſſertion, that the prefent War will produce
fifty years Peace, then indeed, however unjuſt
it might be, there would be fome policy in continuing it. But what reaſon have we to expect
a Peace of fifty years? Whatever may be the
fate of France, does the Hiſtory of Britain or
of Europe juſtify fuch an expectation? At the
beginning of 1792, Mr. Pitt, aſſured us, of fifteen years peace ; at the beginning of 1793, we
found ourſelves plunged into a moſt expenſive
and alarming War! Is Mr. Young a better pro-
phet than Mr. Pitt? A little more than a year
before the commencement of the prefent War,
Mr. Pitt was defirous of involving us in hoftilities with Ruſſia, on account of her aggrandizement, and is not the fame caufe of quarrel likely
to exiſt, in an encreafed degree, whenever our
difpute with France ſhall conclude? Nay, as
all the powers in Europe are only ſmothering
their

their old rooted animofities, in order to join in
a common caufe, will they not quarrel in cafe
of a failure in their prefent project, by blaming
each other with the fault ; or in cafe of fuccefs,
will they not quarrel about the fpoil ? If Auftria
does not gain fomething confiderable, either in
France or elfewhere, can it be fuppofed fhe will
allow Ruffia and Pruffia quietly to retain what
they have taken of Poland ? In which ever
point of view we look at the prefent attempt
upon France, either of failure or fuccefs, the
refult is more likely to leave the feeds of future
Wars, than the profpect of any former ftruggle
in Europe. The Emprefs of Ruffia is perfectly
well aware of this, therefore, fhe allows the
other powers to wafte themfelves, and referves
her ftrength till the day of reckoning fhall
arrive *.

But it would be endlefs to expofe the folly and
injuftice of the general motives for continuing
the prefent War : It is of moft real importance to know, what Adminiftration defign by
it : And here, if we are to be guided by their

---

* This cunning Princefs had fucceeded in perfuading the
late King of Sweden, to take an active part in the prefent
crufade, and it is tolerably well known, that her defign was
to have feized his dominions, when he had fo far exhaufted
himfelf, as to be incapable of refiftance. This is the reafon
why Sweden and Denmark, now obferve a ftrict neutrality,
and keep a watchful eye over her. The King of Pruffia has
been drawn into the fame fnare that was laid for the late King
of Sweden, but it is generally believed, that the purfe of
another power, as well as the affurances of the Emprefs, was,
in the *very firft inftance*, employed to prevail on him. The
confcioufnefs of his own danger from the Emprefs, now
makes him fo unwilling to wafte his troops againft France.

declarations,

declarations, it is but honeſt to confeſs total ignorance, for it would ſeem they do not know
themſelves. But it is not from the unintelli-
gible hypocritical memorials of Miniſters, that
the deſigns of a Government are to be diſ-
covered, ſo much as from the general tenor of
their conduct.

The Engliſh Court profeſſedly began the
War, in conſequence of the decree of fraternity, the attempt to open the Scheldt, and the
aggrandizement of France. When ſatisfaction
on all theſe points was obtained, and France was
ſo far humbled, that ſhe would have agreed to
any reaſonable terms of accommodation, Mr.
Fox, in the month of June, moved an Addreſs
in favour of Peace, as the avowed cauſes of War
no longer exiſted. To this, Mr. Pitt objected,
on the ground, that, although certain points
might be the occaſion of a War, yet in its conduct and events, there was no ſaying what objects might ariſe, which it might be prudent to
obtain; that, therefore, though the original
grounds of the War no longer exiſted, yet his
Majeſty's Miniſters thought it would be proper
to continue hoſtilities, in order to procure indemnity for the paſt, and ſecurity for the future.
As to theſe two words *indemnity* and *ſecurity*,
while it remains undefined, what indemnity
and ſecurity are deſired, as is the caſe at
preſent, they have no meaning at all, becauſe,
they may be made to mean every thing, or any
thing, or nothing: They may be interpreted
juſt as the Miniſter pleaſes: He may deem a
promiſe to pay a certain ſum, as was the caſe
in the late diſpute with Spain, a ſufficient indemnity, and an aſſurance of faith, though no
more

more to be depended on than the faith of the Emprefs and King of Pruffia, fufficient fecurity: Or he may deem nothing lefs than the complete conqueft of France, to be fufficient indemnity and fecurity. To fight for thefe two objects, unlefs they are. precifely defined, is to fight on the moft blind fpeculation.

Fighting for new objects, that may arife in the conduct and events of a War, is nearly the fame thing, as fighting for undefined indemnity and fecurity ; and in fpeaking of the one, the other may always be underftood to be implied. They are both equally the Minifter's fpeculations ; for how can the people know when the Minifter will think he has obtained fufficient indemnity and fecurity, or when he may think it proper to defift from attempting new objects, that may arife in the conduct and events of a War ? The prefent War was profeffedly begun for certain fpecified objects, but in its conduct and events, thofe original objects being obtained, we are next to continue hoftilities for undefined indemnity and fecurity, and then we are to eftablifh Monarchy in France !—The conduct and events of the War may yet give rife to new objects, and our wife Minifters may think they have not indemnity and fecurity, unlefs they not only eftablifh fuch a government in France as may pleafe them, but alfo feize fome of her territory.—The conduct and events of the War may again give rife to new objects, and indemnity and fecurity may be thought to require, that we fhould deftroy the government we had juft given to France, as the King of Pruffia did that which he recently
guaranteed

guaranteed to Poland, and we may agree to divide the whole French territory among the allies.—Still, in the fame manner, new objects arifing, may carry us further, and it may be thought neceffary, for indemnity and fecurity, to obey the exhortations of fome of our divines, and to exterminate the whole French people. —Nay, if fuccefs attends us, why fhould we ftop fhort with France ? New objects may arife in the conduct and events of War, which may, according to Mr. Pitt's reafoning, juftify its continuance, in hopes of the conqueft of the whole world !

But, although from their declarations, we cannot precifely afcertain what are the objects of Minifters in the prefent War, we muft conjecture from their conduct, that they are much beyond what it is thought prudent to avow. Why was our intention of giving a monarchical form of government to France concealed, till the fur-render of Toulon ? Had any new circumftances occurred in France, which made it more ne-ceffary to interfere in her internal concerns, in Auguft, when Lord Hood took poffeffion of that town, than in February, when the War was begun ?—No. But in February, the peo-ple were not ripe for approving fuch a project, therefore, it was concealed from them: Nor are they now ripe for approving, what in the conduct and events of this War it may be thought proper to attempt. The ultimate objects muft be brought forward by degrees, otherwife they might, perhaps, ftartle the nation.

Since

Since not only the Britifh Government, but
the allies in general, appear, by their inconfiftent
conduct, to conceal the true objects of the War ;
fince, by taking Valenciennes for the Emperor,
Toulon for Louis the XVIIth, and fummoning
Dunkirk to furrender to the King of Britain *,
there appears to be fomething of felfifhnefs,
under all the plaufible difinterefted profeffions
of giving happinefs to France, how are we to
folve this myftery ? How are we to be guided
in fearching for the truth ? Are we not to
judge both of men and governments, rather by
their actions, than their profeffions ? If, under
pretence of improving an eftate, an attorney
had taken poffeffion, and by dint of law had
wrung it from the legal proprietor ; if the
fame attorney were to offer himfelf to another
man, who knew of this tranfaction, and to fay
to him, "your eftate is much deranged ; your
" grounds are neglected, and your tenants idle,
" diffipated, and wicked ; put your eftate into
" my hands for a fhort time, I will make it
" productive and beautiful, and your tenants
" induftrious, virtuous, and happy :—it will,
" indeed, coft me many thoufands of pounds,—
" perhaps half my fortune ; but I am refolved
" to do all for your benefit, though I owe you
" no gratitude, and expect no return."—If a
man who was known to have acted fo trea-
cheroufly in one inftance were to attempt to

---

* The Prince of Saxe Cobourg's proclamations, in April
laft, in the firft of which he engaged to reftore the Conftitution of 1789, and in the fecond, recalled that promife,
fhould alfo here be recollected.

repeat his villany, would he be believed or trufted ?—Would he not rather be kicked out of the houfe of him to whom he made the offer ?—If, then, it is from the actions of the man that we form our notions of his character and defigns, rather than from profeffions which we know him moft fcandaloufly to have belied, why fhall we not form our opinion of the character and defigns of a Government in the fame manner ?—Have not Pruffia and Ruffia robbed Poland, under the pretence of giving her a good Government, and making her happy ?—And are they not now holding out the fame pretences to France, in hopes of getting poffeffion, and plundering in the fame man-ner, that nation ?—Has a fingle power in Eu-rope remonftrated againft the robbery of Po-land, as it was their duty, and more particu-larly the duty of Britain, which affected to be alarmed at the ceffion of Oczakow ?—Is not filence, in this cafe, confent ? How then are we to guefs at the objects of the prefent War againft France, but by looking at what has been done in Poland, fince the fame powers are combined againft the one, which actually robbed, or tacitly confented to the robbery of the other ?—which confented to the difmem-berment of Poland, which had neither inter-fered in the internal concerns of other powers, nor infringed treaties, nor violated the rights of nations, nor aggrandized herfelf by con-queft ; whofe new Conftitution was approved by Mr. Burke, and the other gentlemen in this country, who were the moft implacable enemies' of French principles, and even fanctioned by

its

its neighbour, the King of Pruffia, who afterwards made it the pretext for his robbery.

Such is the prelude to what, in common
fenfe, we muft conclude, is defigned to be acted
in France. If bribery can produce treachery
and civil commotion, which, feconded by external force,—if, in fhort, by any means,—for
the laws of nations and humanity are laughed
at,—the allies can conquer France, they will
treat it as they have done Poland, and difmem-
ber it in fuch a manner that it never may
again lift its head among nations. They will
give it fome puppet for a monarch, and, under
pretence of awing Jacobinifm, will keep up a
large ftanding army, for which France will be
obliged to pay. This done, the balance of
power, for which we have fquandered fo many
millions, will be compleatly annihilated; and
if Ruffia, Auftria, and Pruffia can agree about
the divifion of the fpoil, they may divide all
Europe among themfelves *. They may make
a fecond partition of France, as they lately have
done of Poland;—they may do juft as they
pleafe all over the continent of Chriftian Europe, for there will be no power able to oppofe
them.—And even Britain will not long be fuffered to retain her independence, when the
navies of France, Spain, Holland, Sweden,
and Ruffia, can be turned againft her;—nay,
they will probably make her the moft exem-
plary inftance of their vengeance, becaufe fhe
has been the nurfe of thofe principles, againft
which, in France, they are now making War.

* For a moft excellent view of this fubject, fee the Letters of the Calm Obferver.

Such,

Such, I believe, are the real objects of the present War; and, if the allies are succesful, they will, in due time, be unfolded; but whether defeat or triumph follows their arms, the ultimate consequences must be equally pernicious to Britain. For, if their true objects are gained by the conquest of France, Holland and Britain will immediately be at their mercy, and we have seen in Poland what their mercy is: If, on the other hand, the allies fail, we shall be obliged to sit down in disgrace a few years hence; and the large additional burthens may provoke a disappointed and aggrieved people not only to make Reforms, but dangerous innovations. In Lord Cornwallis's late peace, the preserving of Seringapatam was justified on the policy of supporting the balance of power in India; but how much more strongly does the same reasoning apply to the preservation of France? Shall our *gratitude* to Austria and Prussia induce us to ruin ourselves?—A gratitude, which is by no means due to them, as it was not for Holland but for themselves they fought. The wisest policy for Britain, therefore, is to follow the example of the Empress of Russia, and rather strengthen than waste herself, now she has gained all she desired when entering into the contest. To suppose that Prussia, Austria, and Russia design, by this War, to give happiness to France, is so truly ridiculous, that it does not deserve a serious comment;—read their own recent histories.

The duration of the War, or the events that may take place during its continuance, it is vain to calculate. It was positively asserted of of the American War, at its commencement, that it would be finished in one campaign ; and last February, we were taught to expect the same thing of the rupture now existing. Wars of a nature like the present, though even of less importance, have, however, lasted to a period, which might excite ridicule if the same duration were predicted of the struggle now making in France. Wars in former times have been continued in that nation forty years.—The War by which Holland was enabled at last to throw off the Spanish yoke lasted sixty years, although Spain was then the most formidable and Holland the most insignificant country in Europe. Mr. Young by predicting, that the present war will bring us fifty years peace, and that every year's War will bring ten years peace in its train, calculates its continuance at five years. On this point, he is much more honest than those who have written on the same side, for they have, and always will, assure us, that another campaign will settle it ; and they will hold this language, even after the experience of twenty years, shall have twenty times confuted them. It is impossible to calculate the duration of the War ; but, while the allies make the subjection of France their object, it must wear the appearance of a long, bloody, cruel, expensive, ruinous contest. It is not that the Duke of York takes Valenciennes, or Lord Hood, Toulon, or Lord Howe, Brest and Bourdeaux, or the Prince of Saxe Cobourg, Rouen, Lyons, and Paris : The greatest part of France, may be conquered by treachery and force ; but will it long remain so ? Can all the

powers in Europe keep a ſtanding army, liable
to imbibe the principles of Liberty, ſufficient to
awe France into perpetual ſubjection ? Will not
the inhuman conduct of Pruſſians and Auſtrians,
which has already ſhewn itſelf in Alſace and
Lorraine, provoke frequent and formidable in‑
ſurrections ? The example of this country ſhews,
that where the principles of Liberty are ſown,
force will rather nouriſh than deſtroy them,
and as the principles of Liberty are firmly
implanted in the breaſts of the French peo‑
ple, they never can be rooted out. Tem‑
porary calamity may diſguſt them at their Go‑
vernors, but they never have been, nor never
will be diſguſted at their principles. France
may, indeed, be apparently overcome, and
Peace eſtabliſhed ; but a people, filled with
high notions of Freedom, never will long ſub‑
mit to Deſpotic ſway : Frequent Wars will
occur, until at laſt, the conquerors will be ex‑
hauſted, and Liberty will triumph. Mr. Young
infers, becauſe the Engliſh Republican Spirit,
in the laſt century ended in Deſpotiſm, that
therefore, the Republican Spirit in France will
end in Deſpotiſm alſo :—Deſpotiſm, may poſ‑
ſibly, ſucceed the preſent Republic, and reign
for a while as it did in England ; but did not
the ſame principles which brought Charles the
Iſt. to the block, alſo expel James the IId. and
bring about the Revolution of 1688 ? And will
not the doctrines now ſown in France, ulti‑
mately ſettle ſome form of Government,
whether Monarchical or Republican, founded
on the principles of Liberty ? But it is ex‑
tremely improbable, that a great and powerful
nation

nation of Enthufiafts will be overcome, even·
by treachery and force. Their ftrength muft
not be calculated by a narrow Court Policy ;
nor, becaufe Heffians, Hanoverians, and Sar-
dinians, will not fight without money, in what
is deemed a common caufe, muft it be concluded, that Frenchmen are equally mercenary :
There is a National Treafury, more powerful
than all the tax offices in Britain, in the breafts
of Frenchmen :—a love of Liberty. The American paper was lower than even that of France
has been, yet America triumphed ; and though
their Government has excited our contempt and,
horror, yet it cannot be denied that the French
troops have lately difplayed energy, enterprize, and
bravery, fcarcely equalled, certainly never furpaff-
ed in the world. Whatever fond hopes may be
formed of the next campaign, I fear they will prove
illufive, for our future fuccefs is not to be calculated
by the events of laft fummer. Another Dumourier, may not be found, to deftroy the
principal army, and leave the northern frontier
unprotected : On the contrary, we have feen
the ruinous effects of his treachery repaired, and
the tide again turned againft us : Nor will his,
treatment encourage further treachery, or the
treatment of the Toulonefe, encourage Roy-
alifm. It may even be feen in the flighting
treatment of the Ex-Princes of France, that
the Allies do not defign to reftore them to their
former fortunes, but that they intend fomething
both againft them, and the Nation at large,
which, it is feared, thofe noblemen will not
agree to, and therefore, inftead of being held
up as confpicuous leaders in what is profeffed

to

to be principally their own caufe, they are kept
in the back ground, and treated with coldnefs.

The fubjugation and partition of France, together with the eftablifhment of an impotent
Defpotifm, being therefore, the evident objects * of the continental Triumvirate, it may
be ufeful to enquire whether Great-Britain will
affift them to the full extent of their views ;
how fhe can ftop fhort of them, and make
Peace ; whether, in any ftage of the War our
Government will be difpofed, of itfelf, to put
a period to hoftilities ; whether, it will not be
obliged to do fo by the remonftrances of the
people ; in what manner it can make Peace,
and what may be the confequences of being
compelled to make it by the public.

The firft of thefe enquiries need be but fhort ;
it is but to read the Treaty with the King of
Sardinia, wherein, it is agreed, to pay him
200,000l. per annum, during the *whole courfe
of the War*, and the other Treaties with Ruffia
and Pruffia, through which we guarantee the
dominions of all the belligerent powers againft
the arms of France. By thefe Treaties, we per-
ceive, that as long as the French poffefs a fingle

---

* It has been faid, that the ceffion of the ftrong holds on
the northern frontier of France, would fatisfy the Allies, and
I have little doubt, that they really would do fo, for the prefent.
It would indeed, be a fmall ceffion of territory, but, it would
in fact, be a ceffion of much more danger to this country,
than, not only the ceffion of Oczakow, but of all Turkey
in Europe. If the Allies poffeffed that bulwark, they might
not only confine the French, while they plundered Holland,
Denmark, and Sweden, and as they have done Poland, but
the whole of France would be laid open to their maurauding
incurfions.

incl

inch of ground belonging to Auſtria, Sardinia,
or any of the allies ; or, as long as the King of
Sardinia thinks proper to carry on the War,
we are bound to carry it on alſo, which is ſimply,
that while any of the allies continues hoſtilities,
we are bound by Treaty to join them, and without a breach of faith, cannot ſtop ſhort, and
deſert their cauſe. If to theſe Treaties, we add
the conduct of our Ambaſſadors in Denmark,
Sweden, Tuſcany, and Genoa, we may very
reaſonably conclude, that the Britiſh Cabinet is
not only embarked to the full extent of the views
of the Princes on the Continent, but is one of
the moſt zealous, and even furious of the allies;
for our Government ſeems eager to ſur-
paſs in violence all that has been done by other
powers. Peace, therefore, originating in the
Britiſh Cabinet, muſt be at a very great
diſtance.

The next queſtion is, will the people patiently
ſubmit to a long continuance of the diſtreſſes
which are always brought on by War, and invariably encreaſed by its protraction ? Will they
quietly ſee their blood ſhed, and their enormous
debts doubled, in a vain attempt to give a King
to France, and to aggrandize, by the plunder
of her territories, the ambitious Deſpots of the
continent ? The Hiſtory of the American War,
ſhews, that national diſtreſs will certainly open
the eyes of the people to the folly of Government; and the growing diſſatisfaction at the
preſent War, ſhews, that a campaign or two
more, will make it as unpopular as ever that of
America was. It will daily become more manifeſt, that we can derive no benefit equal to

T

the

the rifk we run, and the actual lofs we muft
fuftain by the continuance of hoftilities. I therefore think the people will, at fome period,
not very diftant, perhaps, put an end to this
War, as they did to that with America, by petitions and addreffes. If then, in one, two,
or three years, the people *demand* a peace, *it
muft be granted.* But how is it to be made ?—
By doing that which Government has fo much
reprobated, by negociating with, and acknow-
ledging the French Republic. It has indeed,
been faid, that Britain may withdraw her forces,
and by fome underhand means, procure a fe-
ceffion of hoftilities, which, until a regular
Government is fettled, will be equal to a
formal Peace : But when the French find the
mafs of the Englifh people refolved on a termination of hoftilities, will they not infift upon an
avowed negociation with, and an acknowledg-
ment of the Republic ? When they find the
Court of St. James's, unable longer to carry on
the War, will not they infift upon their own
terms ? And will not a humiliating compliance,
as was the cafe with America, be the confequence ? It is with this War, as it is with Parliamentary Reform : At prefent, Peace and Reform, might be made with the greateft advantage, in the fame manner as a Peace with America, long after the commencement of that unfortunate conteft, might have been made on
beneficial terms. But obftinacy has of late been
the characteriftic of our Government. After
the firft campaign with America, fhe might have
been reconciled to us ; after the firft campaign
with France, we have it in our power amply to
obtain

obtain reparation for all which gave us offence.
It is ridiculous to talk of the inhumanity of negociating with the prefent rulers of France,
when we recollect, that we in 1777, negociated
with, and made *dear Allies*, of the wild favages
in America, and inftigated them to make War
upon the United States, which they did in the
moft horrible manner : it is ridiculous to talk of
our dignity being infulted by negociating with
the rulers of France, when the ignominious
treatment of the Reprefentative of our King,
by the Ottoman Porte, is recollected * : it is
ridiculous to fay, we can have no fecurity for
the continuance of Peace, as the rulers of France
may daily be fupplanted, for, no Government
in Europe, obferves Peace longer than it is its
intereft to do fo, and without expatiating on the
late want of faith in Ruffia, Pruffia, &c. with
regard to Poland, I will venture to affirm, what
is certainly true, that every party which has
governed France, during the laft four years, and
every party likely to fucceed to the Government, would, has been, and will be defirous,
and even proud of keeping Peace with Britain.
But of late, our Court has commonly perfifted

---

* Before our Ambaffador is introduced to the Grand
Seignor, he is obliged to eat fome food. which is given him
in the Palace, and to put on a cloak, worth about 30l. prefented to him at the fame time. When he comes into the
fublime Prefence, he is held by the arms by two officers, who
will not permit him to bow of his own accord, but who, laying their hands on his head, *force* him to bow. They then fay
to their Sovereign, " Here is a poor man We found him
" hungry, and we fed him ; we found him naked, and we
" cloathed him."—Where is the *dignity* of the King's Reprefentative on this occafion ?

to

to the laft extremity ; till the popular tide has
rifen to fuch a height that it was forced to un-
conditional fubmiffion. And, I fear, the prefent
conteft will be profecuted, till Peace muft be
made on *any* terms, and then Mr. Burke, may
do what he ridiculed in Lord North's Adminiftration, towards the conclufion of the American
War, and fay to the French, " Now do have
" a King * !''

The evident determination of the Government to profecute the War, is not fo dejecting
a circumftance, as the too general encouragement given, not only to panegyrics on the corruptions and defects of our Conftitution, but to
the moft falfe accufations, and the moft fhame-
ful calumny againft thofe who defire the reftoration of tranquility. To petition for Peace,
is deemed fedition ; to contend in Parliament
for Peace, is deemed treafon, for what elfe can
we conclude, from the abufe thrown on the
Glafgow petition, and the infinuations of Mr.
Powis †, refpecting Mr. Fox. The many wicked
afperfions thrown on the character and conduct
of the latter of thofe gentlemen, may, however,
be juftly conftrued into eulogiums on his public

---

* This, indeed, has already been faid in Mr. Pitt's late
extraordinary Manifefto.

† This gentleman, in the Houfe of Commons, after be-
ftowing many eulogiums on Mr. Fox, faid, though he was
perfectly convinced of the Right Hon. Gentleman's integrity, yet his conduct in that Houfe, was exactly fuch as
an advocate of the French Convention would purfue.
This, out of doors, was turned into a downright affertion as
a fact ; and Mr. F. was even reprefented in the print fhops,
as the advocate of the French, with a brief and a fee in his
hand.

virtue ;

virtue; for, if ever there was a man, who con-
scious of acting with rectitude, maintained the
true interests of his country with firmness, con-
sistency, and moderation, against all that could
deject and terrify, he is the man : If ever private interest and public fame ; if ever the sweets
of social life, and the prospects of state eleva-
tion, were sacrificed to the national welfare,
and to the liberties and happiness of mankind,
they were sacrificed by Mr. Fox, last winter.
He stood forward, almost alone, and with gi-
gantic power, arrested the Government in its
wanton intoxicated career.. Sedition and insurrection had been declared to exist, the Attorney-
General's table was said to be loaded with hun-
dreds of indictments, and thousands were recorded as disaffected persons, upon the authority
of anonymous letters*, and the veracity of common informers. The suspension of the Habeas
Corpus Act was announced in the House of
Commons, and had it taken place, it is probable, that every man who presumed to enquire
into the propriety of the measures of Government, who praised Liberty in England, or who
dared look cheerful when events occurred favourable to it in other countries, would have
been dragged to a dungeon. But Mr. Fox, stood
forward with truth and energy. The Government was awed :—It paused :—and, finding the
proofs of insurrection and sedition vague and
trifling, it refrained from measures, which, the
delusion of the people might then, indeed, have

* This was done by Mr. Reeves. See Mr. Law's letter
on his secession, from the Crown and Anchor Committee. *

applauded,

applauded, but which their fober reafon, at an
after period, muft have condemned and execrated. He undauntedly ftruggled to avert the
calamities of War ; he did not fucceed : But
he fucceeded in what was of much more immediate importance perhaps, in fhielding the remaining Liberties of the Englifh people.

His enemies have alledged motives for his
conduct, not only bafe, but incredible. The
Tory Jacobins, have accufed him of being the
hired advocate of France ; the Revolutionary
Jacobins, with being fpurred on only by a felfifh
ambition, and the moderate, honeft alarmifts,
though agreeing with neither of thefe, were
not inclined to attribute virtuous motives to a
man, who not only differed from them in political fentiment, but whofe character they found
equally befpattered by the extremes of both
parties.

The accufation of the Tory Jacobins, is evi-
dently fo unfounded, that it is unneceffary to
wafte time in refuting it. The ruling Powers
of France have changed fo often, that fuch a
thing, had it exifted, muft long ago have been
difcovered. Nay, one of the charges againft
the Briffot party, made by the party of the
Mountain, is, that they too precipitately involved France in a War with Britain ; and even
the fupporters of the War at home, affirm, it
was unprovoked on our part, and that the
French were anxious to commence it. Briffot's,
was the ruling party laft winter, and if it was
true, that they were eager for hoftilities, is it
probable, they would bribe any man to prevent
them ?

them ? Nor could the detraction of the ruling men * in the Convention, at the moment the rupture was made, be conftrued, as a proof, that Mr. Fox was connected with them. But, laft winter, paffion had fo blinded the moft alarmed of the alarmifts, that the moft palpably unfounded affertion, if agreeing with their wifhes, received the moft implicit credit : For, " there are feafons of believing, as well as dif- " believing : And, believing was then fo much " in feafon, that improbabilities, or incon- " fiftencies, were little confidered. Nor was " it fafe fo much as to make reflections on " them. That was called, the *blafting of the* " *plot*, and difparaging the King's evidence †."

The conduct of many parliamentary men gives fome colour of truth to the charge of the Revolutionary Jacobins. It is an incontrovertible fact, that oppofition to Government, has too often arifen only in the hope of gratifying a perfonal ambition ; but it is alfo a fact, ftill more incontrovertible, and a fair examination of circumftances, will clearly fhew, that Mr. Fox, laft winter, could not be actuated by any motives of that nature ; but, that on the contrary, his conduct was the very laft which would have been purfued, either by an avaricious, or by an ambitious man : For, if his prime object had been a place, or a penfion, there never was a more favourable opportunity

---

* See Kerfaint's Report, wherein Mr. Fox, is much more çalumniated, than Mr. Pitt.

† See Bifhop Burnet's account of the Alarmifts, at the time of the pretended Popifh Plot, in his Hiftory of his own Times. Vol. I. P. 448.

for gaining it. To oppofe Government was then generally deteftable, almoft dangerous: to defert principles and parties, and coalefce with Adminiftration, was deemed by the country, the height of public virtue. It was honourable in the extreme, for Whigs and Tories to embrace and co-operate: Confiftency of conduct became a crime, and apoftacy the pureft patriotifm. The people confidered it Mr. Fox's duty to join the Court party; his acceptance of an official fituation, and thereby gratifying avarice, and even ambition would have received the warm applaufe and gratitude of his countrymen, as a facrifice of party views and perfonal antipathies, to what they believed the national welfare.

Such were the temptations to induce Mr. Fox to follow his own private intereft, and indulge his ambition. If, on the other hand, we view the reafons he had to *deter* him from fupporting Peace and Reform, we fhall find them not only inconfiftent with avarice and ambition; not only that he was to forego all hopes of fharing the honours and emoluments of office; not only that he was to incur the refentment and odium of the nation, and be branded as the leader of fedition, on the one fide, and accufed of pufilanimity and infincerity, by thofe, among whom he was claffed, on the other; but his deareft friends, and moft valuable connections, were to defert and revile him with a malignity, and injuftice, which his oldeft enemies never could arrive at. Of about two hundred coadjutors in the Houfe of Commons, fcarce fifty
adhered

adhered to him ;—of about one hundred in the House of Peers, there remained not more than fix !—Inftead of being courted and adored, to be fhunned and calumniated, by an hoft of men of the greateft fortunes and talents in the kingdom, was furely no encouragement either to avarice, or ambition. To have all claffes, to have Tories and Whigs, to have thofe who are called Jacobins, and thofe who were called Friends, join in the outcry againft him ; to lofe both popularity and court favour, and even the the enjoyments of private fociety ; to encounter at once, the frowns of the throne, and the indignation of the people, required courage, independence and abilities rarely to be met with. The undaunted, difinterefted exertions of Mr. Fox laft winter, in favour of Freedom, expiate all his former errors. Indeed, a recollection of circumftances gives reafon to hope he never will again be betrayed into fuch errors as thofe which fome years ago rendered him moft unpopular. For if we may believe report, the moft odious of thofe meafures were prompted and executed by perfons who have fince betrayed and deferted him. The coalition in 1783 was projected by a noble deferter, now at the head of the law department, and the negociation was carried on, and the meafure enforced by him and his train of alarmifts. The Eaft-India bill, which begot the *charter alarmifts*, and at once pufhed Mr. Fox from power and popularity, was the production of Mr. Burke, whofe meafures Mr. Fox found it always more eafy to fupport in public, than oppofe in private. To the dog

U

matick

matick opinions of that gentleman, therefore, the miſtakes of Mr. Fox are greatly, if not wholly, to be attributed, and ſo well aware is Mr. Pitt, of Mr. Burke's unruly temper, that it is not probable, he ever will conſent to his admiſſion into the cabinet. If to the coalition and the India biil, is added, the ſupport Mr. Fox gave to certain great perſonages, who have alſo deſerted him, then, all that has made him unpopular, may be ſummoned up. It is his perſonal attachments that have injured his public character, and we now find thoſe for whom he has made ſo important a ſacrifice, eagerly aggravating the approbrium which originated in a reſpect for their opinions, and a zeal for their ſervice. But their conduct may prove fortunate for his reputation. Unincumbered by their baneful influence, and following the dictates of his own reaſon, his integrity and wiſdom, muſt ultimately be acknowledged; and though he may ever remain unrewarded with either place or popularity, the purity of his views, and the prudence of his councils, may yet ſave this infatuated country.

It would be endleſs to enumerate the libels that have been, and daily are, publiſhed againſt him. Mr. Young's book contains one of the moſt fallacious and wicked; and it may ſerve as an epitome of the others. In the debate on Parliamentary Reform, on the 7th of laſt May, Mr. Fox ſaid, " *If* the King and the Houſe " of Lords were unneceſſary and uſeleſs branches " of the Conſtitution, let them be diſmiſſed " and aboliſhed; for the people were not made " for

" for them, but they for the people *. If, on
" the contrary, the King and the Houſe of
" Lords were felt and believed by the people,
" *as he was confident they were,* to be not
" only uſeful, but *eſſential parts* of the Con-
" ſtitution, a Houſe of Commons, freely choſen
" by, and ſpeaking the ſentiments of the
" people, would cheriſh and protect both
" within the bounds which the Conſtitution
" had aſſigned them †." Mr. Young artfully
drops the hypotheſis, and throughout ſeveral
pages, accuſes Mr. Fox with recommending the
diſmiſſal of the King, and abolition of the
Houſe of Lords!—He omits the context,
wherein Mr. Fox ſays, he is confident the
people *feel* and *believe* the King and the Houſe
of Lords to be *uſeful* and *eſſential* parts of the
Conſtitution, and that a Houſe of Commons,
freely choſen, and ſpeaking the ſentiments of
the people, would *cheriſh and protect them.*
Could there be a greater tribute of reſpect and
approbation of thoſe two branches of the Conſtitution than was paid by Mr. Fox, in declar-
ing the people love and will protect them?—
Could there be a more groſs miſrepreſentation
and flagitious calumny than the aſſertion of Mr.
Young?

* Mr. Young, in a note on this paſſage, ſays, the Nobility and the King made the people, and that therefore the
people were made for them! Upon his own mode of arguing it may, however, be proved, that the King was made
for the people; for did not the people make the Brunſwick
the Royal Family of this country? Mr. Young, I ſuppoſe,
would have the Engliſh people made *for* the King, as the Heſ-
ſians are made for the Prince of Heſſe:—to be ſold.

‡ See Debrett's Debates.

Deſpiſing

Defpifing the thoufands of atrocious libels, and regardlefs of his own intereft, ftill we fee Mr. Fox, unfubdued by menace or allurement, perfevering with intrepidity and moderation, in fupport of the Peace, Liberties, and Happinefs of his country. But whatever confolation may be found in his conduct, the general view of public affairs is full of dejection and alarm. On the continent, the laws and rights of nations are trampled on with impunity, without remonftrance, and an extenfive dangerous fyftem of robbery is eftablifhed. The balance of power is loft.—almoft forgotten ; and whether a Republic is attempted in France, or a limited Monarchy, like our own, in Poland, the fact of defiring Liberty, *in any degree*, is fo offenfive to the Combination of Defpots, that they inftantly take arms againft it ;—they deem a wifh for Freedom fufficient to juftify all forts of maffacre, devaftation, and plunder. There appears to be no medium ;—no hope of compromife, can at prefent be entertained. An univerfal War is kindled, which threatens the complete annihilation of Liberty on the one fide, or the total deftruction of all eftablifhed Governments on the other ; for fuch appear to be the views of the two parties, accordingly as the fcale of fuccefs inclines in their favour. A permanent and equitable Peace, therefore, can only be expected, after a long, equal conteft, fhall have deftroyed the means of further warfare ; after both parties, in point of conqueft, fhall find themfelves juft where they began, but mutually weakened by bloodfhed

and

and expence; after they shall have exhausted themselves into tranquility.

At home the prospect is not less full of dejection and alarm than on the continent. In Ireland, a most extraordinary bill has passed, to prevent the people from expressing their wishes : —In Scotland, the most *unprecedented* punishments have been inflicted on those who have advised a peaceable and constitutional Reform of Parliament. These new and alarming experiments have been successfully made in the extremities of the Empire, and it would appear that force is preparing, in order to insure their reception, in the interior. Barracks are erecting in every part of England, where a standing army is to be kept, insulated from the people ; and if that is found insufficient for the purposes of Government, foreign Mercenaries may, by treaty, be landed * to overawe and secure submission. Our Constitution so much boasted for its blessings, and its excellence, is libelled with impunity, as corrupt and ugly, by those who support the Government, and the libellers are rewarded with places and pensions for saying, that " extravagant Courts, selfish Ministers, " and corrupt Majorities," are intimately interwoven with the practical freedom of Britain, and *are good*, while those who affirm they do exist and *are bad*, are punished with the pillory,

* The 7th article of the late Hessian Treaty sets forth, that " If it should happen, they" (the Landgrave's Troops) " shall be employed in Great Britain or Ireland, as soon " as the notification, in such case, shall be made to the " Serene Landgrave, they shall be put on the same footing, " in every respect, as the national British Troops."

not for faying they are bad, but for faying they do exift. To call the Conftitution *pure* * has become a crime ;—to call it *corrupt* recommends to minifterial favour. Mr. Young fays " its fpirit and principles admit of torturing " at pleafure" (p. 199) and I fear it is at prefent juft what the Government choofes to make it, —juft what the people will bear. The Freedom of the Prefs is deftroyed by affociators † for its defence ; the facrednefs of the pulpit is brutally attacked by pretended combatants for religion‡; all confidence in public men has received a mortal blow by thofe who have called moft loudly for confidence §, and the profeffed champions of the Conftitution are the moft bufy in " knocking it down." Petitions are treated with contempt ; petitioners ftigmatized as traitors, and the vague charge of fedition, has put a feal on mens lips. We are to be brought back to darknefs and barbarifm ‖

---

* See Mr. Young's remarks on the petition of the Friends of the People

† Mr. Reeves, while he charges publicans to beware of taking in what he calls feditious newfpapers, &c. fays, he wifhes to fupport the true Liberty of the Prefs !

‡ See the account of the treatment of the Rev. V. Knox by the Militia Officers who have taken up arms againft the *Atheifts* of France.

§ By the apoftate in the caufe of Parliamentary Reform.— Mr. Pitt.

‖ Mr. Young not only recommends the abolition of Sunday Schools, and the Liberty of the Prefs, but fays, the poor fhould not be taught to read, left they fhould read fuch dangerous books as Mr. Paine's !—I wonder he did not alfo recommend the cutting out of men's tongues, left they fhould fpeak feditious words. Without tongues they would be eaqully, perhaps more ferviceable as flaves; as hewers of wood and drawers of water.

as the only ftate in which we can be happy, for knowledge in the poor, is found dangerous to the State, and ignorance and intolerance its beft fecurity. There are but two meafures necef-fary to accomplifh all this : Firft, eftablifh a ftanding force, fufficient to intimidate or punifh the refractory ;—Secondly, fupprefs the Free-dom of Speech, and of the Prefs.

But all exhortations in favour of Freedom, are fo generally confidered, at prefent, as wild and delufive theories, that it may not be im-proper to call in the affiftance of Mr. Young's " Experiment," " Practice," and " Events," to fhew what have been the confequences of times fimilar to the prefent. I fhall leave the reader to judge from the following extracts from the fecond vol. of Rapin's Hiftory of England, how far the conclufion of the reign of Charles the IId. refembled the prefent time. But I beg the comparifon may be underftood as relating to the nation at large, and not as applying to his Majefty. With fome changes of words, but none of meaning, I think, the national temper in 1684, and in 1793, will be found to be ftrongly alike.

" From this time, the King, during the reft
" of his life, governed not only without a Par-
" liament, but with an abfolute power. When
" he faw himfelf out of the reach of the Par-
" liament, he entirely threw away the mafk of
" diffimulation, and fhewed, that the Popifh
" Plot, the profecution whereof he had lately
" recommended fo earneftly to the Parliament,
" appeared to him but a mere chimera, or at
" leaft, he did not think it near fo dangerous as
" he

" he would have had it believed. It is neceſſary
" to unfold the cauſes of ſo ſurpriſing a Revo-
" lution. By the artifices of the Court, and
" the natural inclination of many Engliſhmen,
" the kingdom was divided into Whigs and
" Tories. This diviſion was ſo carefully fo-
" mented by the Court, and the Popiſh party,
" that at laſt it became very great. To render
" the two parties irreconcileable, it was inſinu-
" ated to the Epiſcopalians, of whom the ma-
" jority were Tories, that both Church and
" Monarchy were in danger, and that the ſcene
" of *forty-one* * was going to be revived. That
" the Preſbyterians †, under colour of providing
" for the preſervation of Liberty, really in-
" tended the deſtruction of the Church, and
" the introduction of Preſbyterianiſm ‡, in or-
" der to which, they were purſuing the ſame
" courſe they had taken in 1640, and the fol-
" lowing years, by undermining the founda-
" tions of Monarchy, for the more eaſy ſubver-
" ſion of the Church. Theſe inſinuations had
" the greater effect, as what had once hap-
" pened, and whereof, the memory was ſtill
" freſh, might happen again. The Epiſcopa-
" lians, terrified with the proſpect of falling
" into the ſame ſtate, from which they had
" been miraculouſly delivered, conſidered the

* " The ſcene of forty-one," the Commonwealth, terrified
the nation then, in the ſame manner as the ſcene in France now
does.

† For Preſbyterians may now always be underſtood Reformers.

‡ And introduction of Republicaniſm.

" introduction

" introduction of Popery *, with which they
" were alarmed, as a diſtant and uncertain evil,
" and the eſtabliſhment of Preſbyterianiſm †,
" as certain and preſent. It is even very pro-
" bable, that many whoſe paſſions were violent,
" looked upon Popery as the leſs evil. In this
" belief, they threw themſelves, as it were,
" deſperately into the Court Party (p. 723.)

" Addreſſes became ſo much in vogue, that
" the ſmalleſt Corporations feared the reſent-
" ment of the Court, if they neglected to ad-
" dreſs. The King received them all very gra-
" ciouſly, and diſtinguiſhed thoſe who brought
" them with particular marks of his favour.
" The Lord Mayor, Recorder, and ſome others
" of the City of London, waiting on him at
" Windſor, with one of a very contrary nature,
" were denied admittance, and ordered to at-
" tend the Council, at Hampton-Court, where
" they received a reprimand from the Lord
" Chancellor. It was pretended that theſe loyal
" addreſſes, as they were called, expreſſed the
" ſentiments of the people in general, though
" they came but from one of the parties. But
" what may make it preſumed that the King
" did not much depend upon the people, not-
" withſtanding theſe numerous Addreſſes,
" which weekly filled the Gazettes, is, that
" he never after dared to call a Parliament‡.
" If theſe Addreſſes had expreſſed the general
" ſenſe of the people, what could have hin-
" dered the King from calling a Parliament,

* For Popery may always be underſtood abſolute power.
† Republicaniſm.
‡ For calling a Parliament, it may here be underſtood,
calling a *reformed* Parliament.

X

" which,

" which, to judge by thefe Addreffes, muft have
" been devoted to him.

" The King was not fatisfied with difcou-
" raging thofe who would have prefented dif-
" agreeable Addreffes to him, but alfo filenced
" and imprifoned the news-writers, which were
" not of his party, while others had liberty to
" publifh daily invectives againft the Whigs and
" the late Parliament (p. 724.)

" Every man, who was not of the Court
" Party, and a furious Tory, was called a Pref-
" byterian †. The Clergy, particularly diftin-
" guifhed themfelves, by fhewing their attach-
" ment to the principles and maxims of the
" Court. The pulpits refounded with the doc-
" trine of paffive obedience and non-refiftance.
" The Clergy, feemed to make it their bufinefs
" to furrender to the King, all the Liberties
" and Privileges of the fubject. According to
" the principles they preached, no Eaftern
" Monarch was more abfolute than the King of
" England. This doctrine was fupported in
" the Courts of Juftice, by all the Judges and
" Lawyers, to the utmoft of their power. All
" this was followed with numberlefs Petitions
" and Addreffes. Any man's thinking of affo-
" ciating the fubjects againft the King, was fuf-
" ficient, according to the current principles,
" to charge the whole Whig Party as guilty of
" the greateft crime imaginable. Thus, the
" violent Tories, who then prevailed in the
" Corporations, were not fatisfied with perfe-
" cuting the Prefbyterians, but alfo made the
" King an arbitrary and abfolute Monarch, as

† A Republican.

"

" if there had been no other expedient to fave
" the Church of England from the attempts of
" the Prefbyterians.

" Though fupported by the Court and the
" Magiftrates, the Tory Party had the advan-
" tage, the Whigs were not difcouraged, in
" the expectation of caufing fome turns, by in-
" forming the people in pamphlets of their dan-
" ger. This did but exafperate the patrons of
" paffive obedience. They took occafion from
" thence to carry the doctrine fo high, *that*
" *when in the reign of James the IId. reftric-*
" *tions became neceffary, they knew not how*
" *to make them,* and many even perfifted in
" fupporting this doctrine, rather than own
" they had been in the wrong, to carry it to
" fuch a height (p. 725.) In fhort, a kind of
" infatuation feized the kingdom, and one
" Party, inftead of coming to a temper, vio-
" lently embraced whatever was moft contrary
" the other" (p. 726.)

The King having thus far fucceeded, thought
another alarm neceffary, in order to terrify the
people into a more full compliance with his defign, and accordingly, the Rye-Houfe Plot was
fet on foot, by which, " the whole kingdom
" being ftruck with terror, the King believed
" he ought to improve it to the eftablifhment
" of his abfolute power, fo as to have nothing
" to fear from any future oppofition. This was
" by depriving all the Corporations, and confe-
" quently all his fubjects, of their privileges. It
" was not proper to ufe abfolute power, but to
" proceed in a manner more politic and more
" dangerous to the people, by engaging them

" to

" to make a voluntary furrender * of their char-
" ters, in order to receive fuch new ones as the
" King fhould pleafe to grant. For this purpofe,
" Courtiers and Emiffaries were fent to the
" more confiderable Corporations to infpire
" them with terror, and intimate to them, that
" fcarce one could efcape, fhould the King ex-
" ercife ftrict juftice. This chiefly concerned
" the Whigs and Non-Conformifts, for the
" Tories were generally very readily blinded to
" obey the pleafure of the Court. Jeffries, par-
" ticularly diftinguifhed himfelf in his northern
" circuit, at the fummer affizes. He forgot
" nothing capable of terrifying the people, af-
" furing them, that a furrender of their charters
" was the only way to avert the mifchiefs which
" hung over their heads. Other Judges and
" Emiffaries did the fame, and at laft, the larger
" Corporations being thus gained, the leffer
" neceffarily followed. So a fudden and great
" change was feen in England, namely, the
" Englifh nation, without Rights or Privileges,
" but fuch as the King would vouchfafe to grant
" her ; and what is more ftrange, the Englifh
" themfelves furrendered to Charles the IId.
" thofe very Rights and Privileges, which they
" had defended with fo much paffion, or rather
" fury, againft the attempts of Charles the Ift.

" To make the people in fome meafure fully
" fenfible of their new flavery, the King affected
" to mufter his forces, which, from one regi-
" ment of foot, and one troop of horfe guards,

* Mr Young advifes the people of England, to furrender
almoft every Liberty they poffefs.

" (raifed

" (raifed by himfelf, with the murmurs of
" many of his fubjects) were encreafed to four
" thoufand, compleatly trained and effective
" men. It might then be feen, that the Mem-
" bers of Parliament *, who oppofed the raifing,
" or at leaft the eftablifhment of thefe guards,
" were not altogether in the wrong. But the
" zeal of the Tory Party was now arrived to
" fuch a height, that they looked on every
" thing which contributed to render the King
" abfolute, as a fure means to ruin the Whigs,
" and confequently as a triumph for them.
" They prepofteroufly imagined that the Court
" only aimed at the deftruction of that odious
" Party, and was folely labouring for the
" Tories" (p. 734.)

Such were the effects of the pretended plots,
and the unfounded alarms in the reign of
Charles the IId. They fo fuccefsfully induced
the people to furrender their Liberties, that
James the IId. was encouraged afterwards to
attempt the eftablifhment of Defpotifm. A
Revolution then became abfolutely neceffary;
and, thanks to the pufillanimity of that Prince,
it was made without bloodfhed.

The bafe infidious tools of Corruption are endeavouring to delude the nation into the fame
predicament in which it was in 1684. They
have fought for that which men moft value,
and they find it to be " PROPERTY." In order, therefore, to deter him from overthrow-
ing the pernicious fyftem in which they fatten,

* Here let Mr. M. A. Taylor's oppofition to the eftablifhment of Barracks be remembered.

they

they cry, " Reform will rob you of your Pro-
" perty!"—But thefe are the delufive, treacherous cries of the hyæna, and will betray ultimately into certain ruin. A Parliamentary Reform has been approved, at various periods, by a
majority of men, both in Parliament and out
of it, and even thofe who never fupported
the meafure, have, notwithftanding, indirectly
condemned the prefent conftruction of the
Houfe of Commons, or approved the principles
on which a Reform is demanded *. When
times of affliction and uneafinefs occur, therefore, our defective Reprefentation will be
deemed, and too juftly, I fear, the caufe of
them. A Reform will then be made, not with
caution, and a dread of going too far, as would
be the cafe at prefent, but with indignation and
vengeance. Moderate men will not be liftened
to. The moft wild theorifts will be entrufted
with the work, and inftead of a peaceful, falu-

* The king, in his fpeeches to Parliament, after the
American War, when Reform was fo much agitated, ex-
preffed his defire to fupport the different branches of the
Conftitution, in their *due balance* :—to fupport the *true
fpirit* of the Conftitution, and to ufe his authority for the good
of the people, for which purpofe alone it was given to him.
Thefe fentiments, according to Mr. Young, are dangerous to
the Government. And even Mr. Burke, about the period
alluded to, faid, the King had gone fo far as to recommend
Reform from the Throne.

In addition to the above, Mr. Burke has called our Reprefentation the " flough of flavery:" Mr. Powis, in 1784,
boafted of affembling a little fenate of independant Members
round him, by which he implied, the majority were not
independent. And the Dukes of Portland and Devonfhire,
with many more Peers, figned a proteft in 1777, againft
an increafe of the civil lift, becaufe it was reported, the money
was employed in *corrupting* Parliament.

tary

tary Reform, we fhall, probably, be involved in all the calamities which at prefent torture France.

Mr. Burke, not when he gloried in the eftablifhment of a *Republic* in America, but long after he began to reprobate the eftablifh-ment of a *limited Monarchy* in France, faid [*], " Great difcontents frequently arife in the beft " conftituted Governments, from caufes which " no human wifdom can forefee, and no hu-" man power can prevent. They occur at " uncertain periods, but at periods, which " are not commonly far afunder. Govern-" ments of all kinds are adminiftered only by " men; and great miftakes, tending to inflame " thefe difcontents, may concur. The inde-" cifion of thofe who happen to rule at the " critical time, their fupine neglect, or their " precipitate and ill-judged attention, may ag-" gravate the public misfortunes. In fuch a " ftate of things, the principles now only " fown, will fhoot out, and vegetate in full " luxuriance. In fuch circumftances, the " minds of the people become fore and ul-" cerated. They are put out of humour with " all public men, and all public parties; they " are fatigued with their diffentions; they are " irritated at their coalitions; they are made " eafily to believe (what much pains are taken " to make them believe) that all Oppofitions " are factious, and all Courtiers bafe and fervile. " From their difguft at men, they are foon led " to quarrel with their frame of Government,

" which

" which they prefume gives nourifhment to
" the vices, real or fuppofed, of thofe who
" adminifter in it. Miftaking malignity for
" fagacity, they are foon led to caft off all
" hope from a good adminiftration of affairs,
" and come to think, that all Reformation
" depends, not on a change of actors, but
" upon an alteration in the machinery."

Before the minds of men are fore and ulcerated, and the principles now fown, fhoot out into full luxuriance, let us, therefore, give each part its proper force, and amend and renovate the machinery of the State, while there is no danger that in doing fo it will tumble to pieces. War is the parent of Difcontent, and Difcontent is the nurfe of Revolution. A continuance of hoftilities will produce the times which Mr. Burke defcribes, and then, as in France, it will be too late to Reform. Inftead, therefore, of wafting our blood and treafure to make a King of France, and to give felicity to that nation, let us feize this favourable opportunity to repair and invigorate our own Conftitution; for the only means of promoting and infuring profperity and happinefs to Britain are a fpeedy Peace, and an effectual Parliamentary Reform.

FINIS.

www.ingramcontent.com/pod-product-compliance
Lightning Source LLC
Chambersburg PA
CBHW020551270326
41927CB00006B/800